A CENTURY OF **FLIGHT**

WH S

A CENTURY OF FLIGHT

THE STORY IN PHOTOGRAPHS
OF THE AIRCRAFT, PILOTS,
TECHNOLOGY AND EVENTS
THAT MADE HISTORY

PETER ALMOND

This book was published in 2002 by WH Smith,
Greenbridge Road, Swindon, Wiltshire, SN3 3LD
www.WHSmith.co.uk

This book was conceived and created by Getty Images Publishing Projects
21-31 Woodfield Road, London W9 2BA
charles.merullo@gettyimages.com

A copy of the CIP entry for this book is available from the British Library.

For Getty Images:
Art director: Michael Rand
Design: Tea McAleer
Picture research: Ali Khoja
Editor: Sam Hudson
Proofreading and indexing: Liz Ihre
Special thanks: Slava Katamidze, Elin Hagström,
Mitch Blank and Valerie Zars

Colour separation by @atColor Srl., Milan
Printed and bound by Nuovo Instituto Italiano D'Arti Grafiche, Bergamo, Italy

ISBN 0-9533939-2-5

M 10 9 8 7 6 5 4 3 2 1

CONTENTS PAGE

Of all the inventions of man probably none has had a greater impact over the course of a century than aviation. Air is not our natural element, but to be able to suddenly roam it at will has been to create barely-conceived opportunities - and dangers. One hundred years ago long-distance travel was measured in weeks and months, powerful nations felt completely secure, and the only man-made things in the sky were wind-blown kites and balloons.

What Orville and Wilbur Wright developed in their cycle shop in Dayton, Ohio, and flew at Kitty Hawk, North Carolina, on 17 December, 1903, allowed history to take a fundamental turn. From that moment a revolution was under way that would forever change politics, commerce, communication, science, war, and the importance of technology to a nation's ability to survive and prosper.

We may marvel at the sight of mass air travel heralding the demise of the great, regular ocean passenger liners, at tiny floatplanes opening up the remote interior of Canada, at the devastation of Hiroshima or at helicopter delivery of troops in Vietnam.

Our humanitarian instincts may be sharpened by the knowledge that we can drop food and medicine by air to starving people in Bosnia or Ethiopia, and we appreciate such public services as Australia's Flying Doctor, police helicopters and air ambulances. We also are aware of the growth in size and spectacular designs of airports – cities and nations' front doors – and by how economies and jobs depend on them.

Often the impact of aviation is best illustrated not by a picture of an aircraft, but by, for example, one of a starving child receiving air-delivered food or a grieving family in the ruins of their home, 'collateral damage' from an air strike.

In assessing a century of manned flight we have less need of imagination than with other major human developments, such as ships or trains. With the exception of the earliest balloons, there has been someone with a camera to record almost every aspect of aviation's progress. Some of the first photographs to be reproduced in newspapers were images of aviation experiments; these in turn stimulated pioneers such as the Wrights. Shortly after World War I 35mm cameras such as the Leica made it possible to capture the ever-faster aeroplanes on film and feed the public's love affair with speed and technology. From a military perspective aviation and photography have often justified each other. In 1858, when Gaspard-Felix Tournachon (Nadar) used a balloon to take a series of photographs of Paris, the advantage of aerial photography for information gathering became apparent to all. By 1914 aerial photography was a well-established military requirement understood by both sides.

1903-1913

The Wright Brothers' 1903 Flyer aircraft was the culmination of years of work that in turn built on decades and even centuries of basic aeronautical theorising and experimentation. In 1799 Sir George Cayley, a Briton now regarded as the 'Father of Aeronautics', established the principles of flight with a design that involved fixed wings, a pilot sitting in a fuselage with a combined tailplane of vertical fin and rudder. Ninety years later a German, Otto Lilienthal, took forward Cayley's ideas with a series of 2,500 hang-gliding flights near his home in Berlin. He died in a glider crash in 1896, but his writings, and particularly a number of photographs of him gliding found their way around the world. Among those he influenced was the French-born civil engineer Octave Chanute, who produced a series of gliders and biplane kite structures in America that were ultimately evolved into the Wright Flyer.

Cayley had established the idea of a shaped wing that produced higher air pressure below than above as it moved forward, thus making it rise, and to this Orville and Wilbur Wright added Lilienthal's concept of 'wing warping' lateral control. Where Lilienthal had shifted his body from left to right to make his gliders turn in flight, the Wrights pulled on wires to twist the wing surfaces. They built their first glider in 1900, their second in 1901, and then did their own tests, some using a wind tunnel, before they built their third. It was glider No 3 to which they attached their own four-cylinder, water-cooled engine for that first famous 1903 flight.

Apart from a rather garbled newspaper report and a few photographs that went around the world, however, the Wrights did little to promote the Flyer. After the US government turned them down in 1905 they stopped flying it entirely. It took three years and a Brazilian airship man, Alberto Santos-Dumont, to get another type of craft into the air, the ungainly 14-bis, in France in November 1906. Apart from a Frenchman, Captain Ferdinand Ferber, and an Irish expatriate lawyer, Ernest Archdeacon, few understood the Wrights' idea of 'wing warping'. Flying fever nevertheless started to grow in Europe. Henri Farman finally won the $50,000 Deutsch-Archdeacon Prize for a one-kilometre course in January 1908, but the one minute flight hardly compared with the 39-minute, 25-mile flight the Wrights had done in their Flyer more than two years earlier.

The Wrights finally made the world sit up with a stunning demonstration of the Flyer at the first Le Mans aviation show in France in August 1908. The British had been particularly slow to pick up on aviation, despite the owner of the *Daily Mail* newspaper, Lord Northcliffe, appointing the first air correspondent in 1906, and offering a prize of £1,000 for the first British pilot to fly one mile in a British-built aeroplane. It took an outsider, the big, illiterate, self-promoting Texan Samuel Franklin Cody, to stir up 'air-mindedness' in Britain. His 'British Army Aeroplane No 1', launched from Farnborough Common on 16 October, 1908, made the first powered flight in Britain. But because he wasn't British he didn't win Lord Northcliffe's prize, and when the aerial navigation sub-committee of the Committee on Imperial Defence decided just afterwards that aeroplanes had no military value he was left with no official backing to continue his aviation project.

It was not until Northcliffe offered another £1,000 prize for the first pilot to cross the English Channel that some Britons started to worry. With the only entrants promising to fly French planes from France, not only British pride was at stake but the realisation dawned that attack on the homeland could come from the air. Two Britons set up The Air League to impress Cody's 'air-mindedness' on the British public and politicians. Louis Blériot's Channel crossing on 25 July, 1909, came nevertheless, and Blériot-designed aircraft subsequently made him a fortune.

In America Glenn Curtiss won $10,000 by flying down the Hudson River to New York City. In 1910 the Imperial Russian Flying Corps was founded. Frenchman Henri Fabre performed the first take-off from water in a powered seaplane and American Eugene Ely became the first to fly an aircraft from a ship. Engines, particularly the rotary Gnome, became lighter and more powerful, and in 1911 came the first multi-engined aircraft and the first long-distance flights: London to Paris, Paris to Madrid, New York to St Louis via Chicago. Nearly ten years after the Wright's first flight, however, it was still to the big airships that people looked for most passenger-carrying. Count von Zeppelin's Delag company provided the world's first airline service in 1910, flying between Frankfurt, Baden-Baden and Düsseldorf.

1913-1923

In 1910 an experimental observation flight over British Army manoeuvres on Salisbury Plain evoked strong complaints that it was 'unnecessarily frightening the cavalry's horses'. Italian Lieutenant Giulio Gavotti, of the new Italian Air Flotilla, flew over Turkish troops in Libya in 1911 and dropped grenades on them, but by 1913 the British still had given little serious consideration to how aviation should be used for war. Only one aircraft – the Vickers FB5 Gunbus – was then designed to carry a machine gun.

World War I started with Britain's new Royal Flying Corps and Royal Naval Air Service able to muster only 113 aircraft and six airships for operational use. France had 176 aircraft, Germany 282, Russia 228 and Austria-Hungary 40. Air combat started with observers shooting at each other with pistols and rifles, and then French aviation pioneer Roland Garros came up with the technical breakthrough of metal deflectors shielding an aircraft propeller so that a machine gun could shoot through it. By April 1915, he had shot down five German planes within 16 days and became the first fighter ace of the war. But after he was forced down behind German lines, Dutchman Anthony Fokker, builder of the German Eindecker monoplanes, studied Garros' design, synchronised the machine gun with the engine and, in August 1915, enabled German pilots Max Immelmann and Oswald Boelke to shoot down several Allied aircraft. The next month Germany had control of the air over the Western Front as Allied planes turned away from the 'Fokker scourge.'

Unlike the slaughter of the trenches below, fighting in the air was between individuals, and pilots became household names. Immelmann gave his name to the 'Immelmann turn' which enabled a pilot to outmanoeuvre a pursuer and get on his tail to shoot him down. Manfred von

Richthofen became famous as the Red Baron, his red Fokker triplane helping to strike fear and offer encouragement in the way knights of old used to display their colours in battle. These aces established a heroic reputation for fighter pilots that survives to this day and did much to encourage huge public support for aviation.

The long, cold flights of the bombers allowed much less flamboyant individualism. Initially it was the 50 Zeppelins in Germany's air fleet that caused the greatest concern to the Allies. But by September 1915, British defences forced the airships into retreat. New German Gotha heavy bombers took over the role, 14 of them killing 162 people and injuring 432 in London in one raid in June, 1917. South African General J.C. Smuts not only ordered two squadrons of Sopwith Camel fighters back from France to defend Britain but recommended retaliatory air strikes against German cities and put air power under independent command. If bombers didn't win wars, at least in March and April 1918, during Germany's last desperate offensive, they helped the Allies stave off defeat on the ground, attacking trenches, artillery positions, ammunition dumps and enemy airfields.

The British Prime Minister, David Lloyd George, needing a quick route to victory, agreed to create the Royal Air Force on 1 April, 1918, and sent new Handley Page 0/400 bombers to attack Germany. From America Brigadier-General 'Billy' Mitchell put his ideas for mass air power to the test when, in September, 1918, he led 1,483 Allied aircraft of all types on a raid on the Saint Mihiel Salient. The air power advocates seemed to have made their point, but within a year of the end of the war the fledgling RAF was decimated and Mitchell was struggling to present his case in Washington.

There was still a civilian world to conquer, however, and within days of 11 November, 1918, pilots and aviation engineers started to become businessmen. France encouraged its first airline, the Lignes Aeriennes Latecoere, to open its service with converted Breguet 14 reconnaissance planes from Toulouse to Barcelona on Christmas Day, 1918. In February 1919, Deutsche

Luft-Reederei started a regular passenger-carrying service between Berlin and Weimar. That prompted a consortium of famous French aviation pioneers, including Louis Blériot, Henri Farman, Robert Morane and Louis Renault to start up Cie des Messageries Aériennes (CMA).

But what of the really long routes? Remembering Lord Northcliffe's 1913 prize of £10,000 for the first transatlantic flight, British Captain John Alcock and Lieutenant Arthur Whitten Brown flew a Vickers Vimy bomber from Newfoundland to Ireland in June, 1919. The British airship R-34 made the first two-way Atlantic crossing. Then came the epic flight of Australian brothers Keith and Ross Smith to Australia, and others to Cape Town and India. By 1924 France was linked by air to 12 foreign countries, subsidised and directed by government into non-competitive areas. The Latecoère airline – simply known as 'The Line' to its devoted pilots – began an airmail service between France and South America that brimmed with daring and adventure. If the British had the adventurous fictional exploits of Biggles, then this was the world of Didier Daurat, Jean Mermoz and Antoine de Saint-Exupéry, whose tales of pilots lost in the desert and of murderous tribesmen captivated millions, young and old.

1923-1933

This was America's decade. From a struggling start, by 1925 the big names of US aviation began to establish themselves, with the privatisation of the US mail routes. William Boeing formed Boeing Air Transport, predecessor of United Air Lines, and built the successful two-seat Type 40 passenger plane. The Loughhead brothers built the Lockheed Vega, designed by John Northrop. The Ford Motor Company's famous tri-motor 'Tin Goose' made its first appearance.

The entire world was being connected by air. Two US Army Douglas World Cruiser biplanes went around the world in 1924; US Lieutenant Commander Richard Byrd flew over the North Pole in May 1926 in a prototype Fokker FVIIa-3m; the Dutch airline KLM established a route to the Netherlands East Indies by way of Dum-Dum, outside Calcutta. And then, in 1927, American Charles Lindbergh grasped the $25,000 Orteig Prize, first offered in 1920, for the first direct flight between New York and Paris – twice the distance of Alcock and Brown's 1919 transatlantic flight.

It is still possible to imagine the enormous international public excitement at his achievement. Disaster had struck a succession of earlier competitors, including René Fonck, the top-scoring World War I French fighter ace, whose overloaded three-engine Sikorsky crashed on take-off at New York, killing two of his crew. Charles Nungesser, the third-ranked French war air ace, took off from Paris for New York but was never seen again. Then, on 19 May, the 25-year-old Lindbergh took off from New York in a specially-made Ryan high-winged monoplane. Thousands watched him go and millions more listened with bated breath as the radio reported him crossing St John's, Newfoundland and out across the Atlantic. Battling fatigue, ice and storms, he was seen crossing the Irish coast, then the English Channel, and as his second night of flying approached he found Le Bourget airport in France ablaze with car headlights as thousands of people willed him to land safely.

Lindbergh was the perfect aviation star: handsome, unassuming, courteous. He even paid a visit to Nungesser's distraught mother in France. When he flew his 'Spirit of St Louis' to Croydon airfield near London even more people wanted to see him than in Paris, including the King of England. US President Calvin Coolidge sent a warship to bring him home to an American welcome rarely, if ever, exceeded. Single-handedly, it seemed, Lindbergh transformed American aviation. Money and talent poured into the aircraft industry. Within days of his flight Transcontinental Air Transport (predecessor of TWA) was formed with him appointed chairman of the technical committee in charge of route surveying. By 1930, air traffic in America was twice that of the whole of Europe with just four airlines: TWA, United, Eastern and American.

There was another prize, however, that stirred nationalist ambitions and proved vital to the advancement of aeronautics – the Schneider Trophy, originally offered by the French industrialist Jacques Schneider in 1912 as an annual contest for the fastest sea plane. Interrupted by World War I, restarted in 1919 and brought to a climax in the late 1920s, it was finally retained by Britain in 1931 with a Supermarine S6B floatplane designed by Reginald Mitchell who used what he learned then to design the Spitfire fighter. Another race, the French Coupe Deutsch de la Meurthe for smaller engines, went on to encourage Germany's Messerschmitt Bf 109 fighter.

For all the achievements, however, aviation was still only for a relative handful of people – the brave, the entrepreneurial and the rich. It was a hard, expensive and exhausting experience to be an airline passenger on the long-haul routes. A person wishing to travel with Imperial Airways from London to Karachi in 1929, for instance, paid £130 and took a flight from Croydon to Basle, Switzerland in a 20-seat Armstrong Whitworth Argosy with toilet and baggage space at the rear. Then, to avoid a dispute with Italy, the passenger took a train to Genoa, Italy, followed by a Shorts Calcutta flying boat across the Mediterranean to Alexandria, Egypt, where he changed to a seven-seater de Havilland 66 Hercules biplane in which two seats had been removed to make room for a bar. The passenger might well have needed it, because he would then fly across Arabia, stopping at nights in desert camps fortified against local tribesmen, and finally take a southern route to Karachi, Pakistan, away from the Persian coast where the locals sometimes took potshots at the 'big birds in the sky'. It was such a potshot in Iraq that killed Arthur Elliott, engineer to British pioneer Alan Cobham, as they flew a DH-50 seaplane on a record-breaking flight back from Australia in 1926.

The battle for long-distance air passengers in the early 1930s came down to four main contestants: flying boats; slow, stable, roomy and luxurious airships; the slow but steady big biplanes; or the smaller, but getting bigger, high or low-winged monoplanes. Increasingly it was the last, epitomised by

Hugo Junkers' low-wing metal aircraft - the first of which was seen as long ago as 1913 – that won out. These designs combined with the 'stressed-skin' ideas of fellow-German Adolph Rohrbach – in which the smooth metal surfaces of wings and tail-plane shared the load stresses of the spars and ribs. The Junker-designed planes were lighter, stronger and had less drag, or wind resistance. The low wings also allowed retractable undercarriages, with wheels folding away into a wing made deliberately thick to take the stresses and strains of flight formerly carried by the support spars of longer-winged biplanes. Both innovations made the plane go faster and save on fuel bills.

1933-1943

A lecture by Rohrbach in the United States in 1926 encouraged Boeing aircraft designers to develop the low-wing, Junkers-inspired Boeing 247, which entered service with United Air Lines in March 1933, and is now recognised as the first modern airliner. The 247 carried only ten passengers, but it left all its competitors standing - for a couple of months. On 1 July, 1933, the first Douglas DC-1 was flown in California, then the larger and faster DC-2 and then, on 17 December, 1935, the first prototype DC-3, one of the most famous aircraft in aviation history. While the 247 was the prototype for the British Bristol Blenheim bomber and inspired both the German Junkers Ju 86 and Heinkel He 111 bombers, the DC-3 was faster, had wing flaps for better control, carried twice as many passengers and had greater range. The airlines and military loved it, and over the next 60 years 10,655 were produced in the US, with others built in Japan and under licence in the Soviet Union. DC-3s are still flying somewhere in the world.

By the mid-1930s flying was the thing for public, business and the armed forces. Airports not only expanded and laid down concrete runways but municipalities bought them up with public backing. Hendon's military air shows, outside London, begun in 1920, were as popular as ever. In 1934 the British government opened all military and civil airfields to the public for Empire Day, an annual event. In 1937, however, there were setbacks. In May the great German airship *Hindenburg* burst into flames as it landed in Lakehurst, New Jersey, with the loss of all on board. The horrific destruction was photographed and described live on radio – one of the most famous of all broadcasts – and so shocked the public, following the crash of the British R-101 airship in 1930, that the *Hindenburg* became the last of the great passenger airships. In July Amelia Earhart, probably the world's most famous woman pilot, was lost on a round-the-world trip somewhere in the Pacific.

Elsewhere, the concepts of military air power were really beginning to strike home. A year earlier Italy had overwhelmed the weak defences of Emperor Haile Selassie, of Abyssinia, with massive surprise air attacks. Italian Major-General Giulio Douhet, had written a book, *Command of the Air*, in 1921 arguing that air power would dominate all future battles. Benito Mussolini, Italy's fascist leader, had also seen how the British were able to police Mesopotamia with aircraft in the early 1920s.

Douhet's argument about aviation's strategic dominance persuaded Sir Hugh Trenchard, commander of Britain's Royal Air Force, to focus his scarce budget resources on bombers more than fighters. Prime Minister Stanley Baldwin summed up the British position in 1932 when he told the House of Commons: 'I think it is well for the man in the street to realise that there is no power on earth that can protect him...the bomber will always get through.' In April 1937, the citizens of Guernica, Spain, learned just what that meant when their town was smashed to pieces by wave after wave of Heinkel He 111 and Ju 52 bombers. Pablo Picasso's painting *Guernica* has come to be the representative image of the terror of war from the air.

In September, 1939, the lessons of Abyssinia and Spain were ruthlessly applied to Poland, and then, in 1940, to Holland, Belgium and France by the former Kondor Legion commander in Spain General Wolfram von

Richthofen, cousin of World War I's Red Baron, and by the Luftwaffe's Chief of Staff, Hermann Goering, who had succeeded to the command of the Flying Circus on the death of the Baron. Unlike the strategic ideas of Douhet, Trenchard and Mitchell, however, German concepts of blitzkrieg were essentially tactical, coordinating air strikes with attacks by rapidly moving armoured columns. With a pre-war construction programme restricted to medium and light bombers and fighters to meet this requirement, however, Germany failed to produce significant numbers of long-range heavy bombers to match the destructive power of British and American bomber raids – a major weakness.

In the summer of 1940 nothing but the Spitfires and Hurricanes of RAF Fighter Command stood in the way of Hitler gaining total dominance over Europe. Fortunately for Britain the RAF had just enough of the new fighters and, with help from a new invention, radar, turned the Battle of Britain into Hitler's first major defeat. Hitler himself ordered a switch from raiding airfields to attacking London in September, giving Fighter Command the breather it desperately needed.

For America, the power of carrier-borne aircraft was revealed brutally on 7 December, 1941, the 'day which will live in infamy', when hundreds of Japanese planes attacked the US Fleet at Pearl Harbor. General 'Billy' Mitchell, hero of the Saint Mihiel Salient assault in World War I had warned of this day, but he could never carry the navy with him. He, like Douhet, fervently believed in strategic air power, but he was court-martialled for insubordination in 1925, resigned and eventually died a bitter man in 1936.

Now America set out to honour Mitchell's memory. Congress awarded him a posthumous Medal of Honor, and in its greatest tribute to him, in April 1942, the US government sent its aviation pioneer Lieutenant-Colonel Jimmy Doolittle on a one-way raid on Tokyo in B-25 Mitchell bombers named after the air power advocate. Doolittle's bombers, launched from an aircraft carrier, caused no significant military damage, but America was able to show it had strategic air power to reach and hurt Japan. The following month a US carrier task force, spared from the Pearl Harbor attack, caught a Japanese carrier force in the Coral Sea. In the first sea battle in which hostile ships never saw each other, aircraft from the two groups fought to a draw, with ships sunk on both sides. US Navy pilots finished the job at the Battle of Midway four weeks later, finally ending Japanese superiority in the Pacific by sinking four of its carriers.

1943-1953

Of all the squadrons in the RAF none is so famous as number 617, the Dambusters, partly because of the popularity of the movie about it, with its stirring music by Eric Coates. The 1943 raid against the Mohne, Sorpe and Eder dams in the Ruhr valley is remarkable because of its ingenuity, daring, dedication and precision at a time when thousands of aircrew were being killed in a massive bombing offensive which seemed to lack any real focus or obvious achievement.

But on the night of 16 May, 1943, 19 Lancasters of 617 Squadron each carried a single round 'bouncing bomb' designed by inventor Barnes Wallis, to be dropped at low level, ending 40 feet underwater next to the walls of the dams. Eight of the Lancasters were lost, but the Mohne and Eder were breached, and floods disrupted power to Ruhr factories and killed 1,300 people. The fact that most of the deaths were imported slave workers, and industrial production was only temporarily affected was of less importance to the Allies than that British morale soared, and had an effect probably equal to that of the Doolittle raid on the American public. Raid leader Wing Commander Guy Gibson was awarded the Victoria Cross, and the squadron was reserved for specific precision bomb raids for the rest of the war. Prior to this Bomber Command had given up precision bombing because it was judged too risky by day and at night the targets were too difficult to see. Air Chief Marshal Sir Arthur 'Bomber' Harris had the personality and authority to persuade Churchill to put the bulk of Britain's war resources into a strategic bomber force that would pound German cities to rubble, destroy civilian and military morale and force Hitler to deflect his armaments production into defending the homeland. Harris' force grew steadily, until by March 1943, he had nearly 1,200 aircraft, a total of 62 squadrons, of which 18 were equipped with the new heavy Lancaster bomber. Harris' thousand-bomber raids destroyed cities such as Hamburg, Cologne, Essen, Nuremburg and Berlin. Hamburg, indeed, became the first city to suffer the horrendous phenomenon of firestorm, in which temperatures reached 1,000 degrees Centigrade. In one night some 42,000 German civilians were killed, and 37,000 injured. But the RAF was finding German night fighters were taking an increasingly heavy toll and by March 1944, Bomber Command was no longer in a position to sustain major night offensives.

The US Eighth Air Force also followed the British example and gave up precision daylight raids because of its increasing losses to German fighters. It was not equipped for night bombing. The tables slowly turned, starting with the introduction in August 1943 of the Thunderbolt fighter escort, equipped with long-range fuel tanks. Then in November came the Lightning fighter and finally the P-51 Mustang, which could escort the bombers the whole length of Germany. Finally, the fuel which fed the Luftwaffe's fighters started to dry up under bombing of Germany's oil supplies.

In mid-1944 Allied air power turned not only to supporting the coming D-Day landings in Normandy but to becoming a direct part of army operations through the use of paratroopers and towed gliders. Paratroopers had been first used by the Germans in Norway and Denmark in 1940, and more notably in their invasion of Crete in 1941. Despite the British disaster at Arnhem in1944, the success of Paras earlier in Normandy is still the historical inspiration for both Britain and America's elite troops to use silent parachute drops as the spearhead for attacks.

In a way, the end of World War II came only just in time. Germany's introduction of the V-1 'doodlebug' flying bomb, and then the big V-2 rocket with its unnervingly-silent final descent, were too little, too late. So, too, was its pioneering jet fighter the Messerschmitt 262, which shot down 427 allied aircraft and could have downed many more if Hitler had not insisted on trying to make it a bomber. Although Britain thought it had the world's first jet aircraft in Frank Whittle's Gloster E28/39, which made its maiden flight in May, 1941, Ernst Heinkel had his turbojet He 178 flying in late 1939.

There have been few greater milestones than the atom bomb dropped on Hiroshima in August 1945. Air-delivered nuclear weapons either put an end to the realistic likelihood of global war, or raise the fear that global nuclear annihilation is an ever-present possibility. It depends on one's views about the concept of deterrence and MAD – Mutually Assured Destruction. For the first time in any country air power reigned politically supreme over all other military arms in the United States, and its air force became independent from the army. The nuclear threat certainly did hold the Soviets back from direct military action to take over the whole of Berlin. Locked inside Soviet-controlled East Germany, West Berlin was merely blockaded from June 1948, to September, 1949. The only food, fuel and supplies for the two million people of the Allied zones in the city in that time had to come by air.

For the moment the Soviets had been bested. But dogged determination to learn was not just a western trait. A warning of Russian ingenuity sounded after four B-29s, forced down in the Soviet Far East at the end of the war, were reverse-engineered by Andrei Tupolev and reappeared in 1948 as the Tu-4. In 1949 the Russians exploded their own nuclear device. Then there came the sudden appearance over North Korea in 1951 of the MiG 15, the Soviets' first jet fighter which Mikoyan and Gurevich put together from German swept-wing studies and the naive supply by the British government of two of its latest and most powerful jet engines. An American F-80 Shooting Star fell victim to a MiG in the first combat engagement of two jet planes. America only got back the edge with later versions of the F-86 Sabre.

The world's military was plunging on with ever faster jets – indeed, in October, 1947, US test pilot Captain Charles 'Chuck' Yeager broke the sound barrier in his rocket-powered Bell X-1 at 670mph. But what of the civilian world? Where were the new jetliners? The answer, it appears, had to do with state support. Just as the US government was funding military jet research which allowed aircraft manufacturers to develop civilian planes, so nationalisation of Britain's two main airlines, the British Overseas Airways Corporation (BOAC) and British European Airways (BEA) gave de Havilland the assurance that BOAC's order for eight of its planned Comets had government backing. The Comet 1 made its maiden flight in 1949, and became the first jet airliner in service in May, 1952. Within 20 months three of the aircraft had crashed with the loss of all on board. In October, 1954, a board of inquiry found metal fatigue in the cabin structures, particularly around the windows. The aircraft was grounded.

1953-1963

If there was one major new aviation lesson to be learned from the Korean War it was the usefulness of helicopters, perhaps best remembered by the public as the little bubble-cabin Bell Sioux helicopters featured in the TV comedy series *M*A*S*H*. Instead of long and rough casualty evacuation by road, the helicopters whisked injured soldiers back to expert medical care within minutes. It is estimated that during the war at least 15,000 troops were evacuated by air this way, some 3,000 of whom would almost certainly have died otherwise. Combat rescue helicopters also won the world's admiration, frequently dashing behind enemy lines to rescue downed pilots while 'rescue caps' of ground-strafing aircraft kept enemy forces at bay.

This was the era of the big military jets, Britain's Valiant, Victor and Vulcan V-bombers and America's B-36, B-47 and B-52s of General Curtis Le May's Strategic Air Command – all different, but to the average man in the street all with one thing in common – they were incredibly noisy. US

Air Force officers had a phrase to silence any public criticism of it: 'Sound of freedom, sir!' Increasingly the Soviets had strike aircraft within a few minutes' flying time of NATO targets in West Germany, and they had new, long-range bombers able to reach even the continental United States.

Andrei Tupolev used his Tu-16 and Tu-95 designs to develop jet passenger aircraft at half the cost and in half the time of those in the West, though the Russians got more than a little of their inspiration from that direction. When the 50-seat Tu-104 made its first appearance at London's Heathrow airport with Aeroflot in 1956 it caused a sensation. It was the only turbojet airliner in regular passenger service in the world and it was not until two years later that the redesigned Comet, now known as the Comet 4, came back into service with BOAC.

The Comet 4 made the first jet commercial crossing of the Atlantic in October 1958, but it was already too late. The 65-seater, longer, more powerful and capable than the Comet 1, was never designed for transatlantic crossings and had to refuel in Gander, Newfoundland on its way to New York. Two weeks later Boeing's new 707 made its maiden transatlantic flight direct from New York to Paris, 100 mph faster and with twice as many passengers. Boeing quickly got orders for it from around the world.

Britain, it seemed, was losing its way. The 1957 Duncan Sandys Defence White Paper, written a year after the Suez debacle, was based on the assumption that the country's requirements had shrunk and no longer needed combat aircraft to reach all the way round the world. It advocated the demise of manned aircraft and wider reliance on missiles. Indeed, one of the first things the government did was to accelerate the Blue Streak intermediate nuclear missile project and start the Blue Steel aircraft-launched stand-off nuclear bomb. It also bought Thor land-based missiles from the Americans as a stop gap. The French, in particular, felt humiliated by Suez and determined to build their own independent nuclear Force de Frappe.

The Sandys white paper also had a strong effect on British civil aviation because it announced a drastic cut in financial backing for advanced aerodynamic development. With the first Douglas DC-8, a major new US competitor to the 707, already flying but not yet in service, Britain appeared to have ceded the big passenger jet business to the US. There was only one solution – merger. The great old names of British aviation began to be rolled into the British Aircraft Corporation, echoing the rise of state-owned Aerospatiale in France.

But perhaps the Sandys paper had a point about missiles versus manned aircraft. The Soviets had the Sputnik satellite in 1957, and the first two Intercontinental Ballistic Missile submarines the next year. The US responded with its first two Polaris-firing subs in 1960, along with the first Atlas ICBMs. That year, Gary Powers' U-2 reconnaissance plane was shot down by a new Soviet missile over Sverdlovsk, even though he was flying at a previously-untouchable 65,000 feet. How could the British – or the French – afford to match all that massive expenditure? The answer, for the British at least, was that they couldn't. In 1960, suddenly realising that Soviet missiles were likely to destroy Blue Streak's silos before the missile could be launched, the government scrapped it as a weapons system. It opted instead for the new US Skybolt air-launched stand-off missile which could at least give the V-bomber force a better chance than it was going to have with Blue Steel, since the former could be launched 1,000 miles from target, while the latter's range was 100 miles.

The Cold War intensified. In April, 1961, the Soviets put the first man into space – Yuri Gagarin. In August they put up the Berlin Wall. Then came the 1962 Cuban missile crisis – probably the closest the world has yet come to nuclear war. France, which had exploded its first nuclear bomb in 1960, decided to pour billions more francs into its own ballistic and tactical nuclear systems. Britain, however, had its strategic direction largely determined for it when for technical reasons the US cancelled Skybolt and decided to focus on the submarine-launched Polaris. Cash-strapped Britain had little choice but to follow suit and buy Polaris off the US. At least the warhead and submarines would be British.

The Skybolt sticks in the British political memory as an example of American perfidy. The truth, as usual, is more complicated. Nuclear-powered submarines armed with ICBMs could hide for months in the ocean depths. Aircraft were more vulnerable. Blue Steel was picked up again, but the odds were stacked against aircraft being the main carrier for the nuclear deterrent – for Britain or for anybody else.

1963-1973

There was still the TSR-2, the only new military aircraft project allowed by the British government in missile-minded 1959, meant to replace the V-bomber force and be the key to the RAF's tactical nuclear capability. Publicly revealed in October, 1963, the TSR-2 had a speed of Mach 2.5 and the first practical terrain-following radar, allowing it to fly lower to the ground than any other aircraft and so to deliver its two nuclear weapons under enemy radar. It was, however, enormously expensive and the Labour Party vowed to cancel it if it won the 1964 general election, so the TSR-2 was abandoned in April, 1965. Instead, 50 American F-111 swing-wing bombers, using a concept invented by bouncing-bomb designer Barnes Wallis, were ordered. That order was cancelled three years later, and it was not for another ten years that Britain got anything like the TSR-2 – the Panavia Tornado. The TSR2 is one of the great might-have-beens of British aviation. Its engine was, however, developed for the Concorde.

Prime Minister Harold Wilson also cancelled the HS681 transport plane, going for the US C-130 Hercules instead, which turned out to be as brilliant and as globally ubiquitous as the DC-3. Then the P1154, another unique British product, went to the wall. This was a supersonic vertical short take off and landing (VSTOL) aircraft designed to replace the Royal Navy's Sea Vixens in the next decade. Actually, the navy hated the plane. It could be operated from small ships, so putting in doubt the need for big aircraft carriers like the 50,000-ton next-generation CV01. Navy chiefs pressed to have catapult-launched US Phantom fighters instead. They got them, but it was a hollow victory. With Britain increasingly Euro-centric the government saw no need for big, world-ranging carriers. CV01 was cancelled, and the Phantoms went when the last of the big carriers were scrapped in the mid-1970s. The new 20,000-ton *Invincible*-class flat-tops were not officially listed as aircraft carriers at all, but as 'through-deck cruisers' designed for helicopter-based anti-submarine warfare.

By 1965, US carrier and air forces were in action in Vietnam, including for the first time B-52s, each dropping dozens of 750 and 1,000-pound bombs in the three-year Operation Rolling Thunder. The lasting images of the Vietnam War, however, are again those of helicopters, scores of them, particularly UH-1 'Hueys' dropping and picking up troops in jungle clearings or rice paddies. It was in Vietnam that the concepts of 'air cavalry' were most developed. Just as the US 7th Cavalry traditionally rode to the rescue of settlers attacked by Indians, so US and South Vietnamese troops used helicopters to rush to the aid of units attacked by Viet Cong guerrillas and the infiltrating North Vietnamese Army, and to leapfrog tangled jungle to launch attacks of their own.

'Back in the world', as the US troops in Vietnam used to say, a revolution in civilian mass air travel was under way. America led the charge, both with the introduction of larger jets and with a huge expansion of airports in the early 1960s. Allegheny Airlines initiated a no-booking system in 1960 with offers of a 36 per cent discount. In 1961 Eastern Airlines started a new 'shuttle' service between New York and Washington DC, Newark and Boston. The bigger jets meant many of the smaller, propeller-driven planes became available for shorter routes. In Britain that year British European Airlines started a London-Glasgow night service using older 130-seat Vanguards. Lower fares meant more passengers and that in turn meant bigger ground facilities. TWA's new terminal at Idlewild airport, New York, had the first covered passenger walkways direct to the doors of aircraft.

Britain's big state airlines started to lose their monopolies. In 1961

Freddie Laker's British United Airways was allowed to compete with BEA over the same European routes, and a year later a package tour holiday company, Universal Sky Tours, bought up three old Lockheed Constellations and started to fly holidaymakers from Manchester to Majorca.

For the plane makers the rise of mass air travel was a challenge, a threat to their traditional, technologically-driven culture of bigger and faster. The cream of aeronautical engineering effort had been going into a supersonic transport, and the four great aviation nations – America, Britain, France and Russia – had poured billions into such projects. In reality there was only one that was likely to come to fruition - the Anglo-French Concorde. But in the US Boeing was putting its money on a much slower and more economical 'jumbo jet' able to carry up to 400 passengers more than 5,000 miles. The crunch came in 1969, one of the biggest years in the history of aviation.

In February, the Boeing 747 Jumbo had its maiden flight near Seattle. In March the prototype Concorde 001 took off from Toulouse. In May an RAF Harrier jump jet took off from a coal yard at London's Euston station and flew to a site in Manhattan, New York, in six hours 11 minutes. In July Neil Armstrong became the first human to land on the moon. Were these omens to be read? Were these complementary forms of aviation, or competitive? Did the Harrier flight mean passengers would soon be whisked from city centre to city centre in vertical take-off airliners? Was interplanetary space travel now within reach for the average passenger, perhaps linked to Concorde developments? And more immediately, would airlines buy both Concorde and the 747? The accountants fell for the jumbo. The 100-seat Concorde was fine for the rich and for business people needing to get to New York to sign contracts effectively before they even took off, but not for the mass of travellers to whom economy was more important than time.

1973-1983

The battle over Concorde went on through the 70s, fed as much by British and French nationalistic pride as anything else. Only the state airline carriers BOAC and Air France had ordered Concordes - just 16 between them, nine built - and with the US Congress having cut off funds for the American SST there was a certain amount of pique in 1973 when Pan American and TWA declined to buy. The tremendous noise of Concorde's engines encouraged the New York Port Authority to take legal action to ban the plane, and so its first scheduled flight was not across the Atlantic but from Paris to Rio de Janeiro in 1976.

If Concorde turned out to be a bit of a blind alley for Europe in commercial terms, the Airbus 300 was not. The project, launched at the Paris Air Show in 1969, joined French, German and British industries into a consortium to produce a medium-range aircraft. It went into service with Air France in 1974 and by 1979 Airbus had broken into the American market with 26 per cent of the world-wide market by value.

But Boeing came back with the 747SP, designed to carry some 300 passengers 6,740 miles. Indeed, in 1976 one flew around the world in a record time of 46 hours 26 minutes. And it was a 747 that got massive media attention when it started to be used for piggy-back transporting of the next big thing in aviation, the re-usable Shuttle spacecraft, from its landing site in California to its launch pad at Cape Kennedy in Florida. When things went wrong with jumbos, however, they went very wrong. In 1977 two 747's collided on a runway at Tenerife, Canary Islands, killing 575 – still the worst-ever aircraft disaster. Neither aircraft should have been there at all, but had been diverted from the neighbouring island of Gran Canaria after a terrorist bomb had gone off. The rise of aviation-related terrorism is, in fact, one of the key things historians are likely to remember about the 70's. The international aspects of aviation, the huge costs of aircraft and airports, the vulnerability of hundreds of people packed into a box with wings all combined to provide a powerful magnet to those who sought to draw attention to themselves or their cause.

Although there had been a number of US domestic aircraft hijacked to

Cuba in the 1960s the first major international terrorist incident was the successful hijacking of an El Al 707 from Rome to Algiers in 1968. In 1970, Palestinian gunmen forced three airliners to fly to Dawson's Field, Jordan, where they blew them up after exchanging their hostage passengers for seven Palestinians held by the German, Swiss and British governments. One Palestinian released by the British was hijacker Leila Khaled. In May, 1972, a Sabena jet was hijacked by Black September terrorists at Ben Gurion airport, Israel, the passengers freed only after the plane was stormed by Israeli troops. A few days later came the massacre at Tel Aviv airport, when three Japanese terrorists used automatic weapons and grenades to kill 25 passengers and wound 72.

The fun was going out of aviation. Pilots around the world went on strike to call for tougher anti-hijacking measures. Baggage checks, security scanners and long queues at airports started to become a way of life. But still it went on, the 1973 Arab-Israeli War embittering more people. It was far from turning people off flying, however. President Carter endorsed airline deregulation in the US in 1977, and within days American Airlines was offering 'super saver' fares between New York and the West coast up to 45 per cent lower than standard. In Britain, Freddie Laker successfully concluded a six-year battle for lower fares when he was allowed to start Skytrain, a no-frills service from Gatwick to New York. The big airlines hated him, as they were forced to offer cheaper 'Club Class' fares, but the public didn't, waiting for hours and days for tickets. By 1980 revenue passenger miles had more than doubled, from 286,000 million to 677,000 million.

In military aviation Western technology was getting the upper hand. First, in the Yom Kippur War of 1973, after being surprised by effective Arab surface-to-air missiles, US-supplied Israeli Air Force Phantoms and Skyhawks, and French-supplied Mirages overwhelmed Soviet-supplied Sukhoi and MiG fighters of the Egyptian and Syrian Air Forces. Then years of basic research into aircraft design finally started to produce results. 'Swing-wing' and 'Fly-by-wire' became the in-phrases. First came the

Anglo-German-Italian Tornado multi-role aircraft. Like the American F-111, its wings moved forward and back according to its speed. Its aerodynamics completely voided everything the Wright Brothers and Santos-Dumont knew to be true, making it inherently unstable. If its engine stopped and the power was turned off it would plummet like a stone. In the 1976 F-16 fighter fly-by-wire came into its own, using computer-guided flight controls to work against the aircraft's tendency, inherent in its design, to flip on its side, and to allow many more flight adjustments than a human pilot could make.

By the early 1980s computers and long-range missiles were beginning to compensate for poorer pilots or older aircraft. The British discovered this only too painfully in the 1982 Falklands Campaign, when French-built Super-Etendards of the Argentine Air Force sank several ships of their task force with long-range Exocet anti-ship missiles. The war was a stunning repudiation of Britain's late-60s decision that it didn't need long-range, catapult-launched aircraft any more. It had only the short-range Harrier jump jet for fleet protection, and no long-range airborne early-warning radar. It was lucky for the British that the Argentines didn't have more Exocets and that their attacking planes were at the limit of their range. But the conflict did prove the efficacy of the Sidewinder air-to-air missiles with which the Harriers were armed.

1983-1993

On the face of it the military and civil worlds of aviation were becoming more and more separate. The Cold War was as intense as ever, but you'd hardly know it if you were an airline pilot or passenger focussed almost entirely on ensuring your baggage wasn't lost, that you made your connection to Brussels or wondered what in-flight movie to watch. Thousands of feet below, however, pilots sat on the ends of runways waiting to be attacked, sailors sat in darkened combat control centres on destroyers and practised shooting down enemy aircraft. It was, perhaps, inevitable that the two worlds should collide sooner or later.

In September, 1983, Korean Air Lines flight 007, a Boeing 747 flying from New York to Seoul, strayed into sensitive Soviet airspace near the Pacific island of Sakhalin. Despite being tracked by Soviet radar for more than two hours and being visually identified by two Soviet Su-15 fighters, the airliner was shot down with the loss of 269 lives. The Soviets claimed KAL 007 ignored efforts to contact it, and thought it was a US EC-135 which flew regular electronic-reconnaissance (ie long-range spying) missions near the route.

The Americans were outraged, but they did a similar thing in 1988 when the USS *Vincennes* shot down an Iranian Airbus over the Persian Gulf, claiming she mistook it for an Iranian F-14 fighter bent on attacking her. This time 290 civilians died. An investigation revealed that in spite of the *Vincennes* having the most advanced aircraft detection, identification and prioritisation system in the world the operators, having just had a surface engagement with rogue Iranian gunboats, the crew thought everything around them was hostile. Nobody remembered to look up Iran Airlines' published flight schedules from Bandar Abbas.

But who was the enemy? Increasingly, not the Warsaw Pact. Iranian-backed terrorists hijacked a TWA Boeing 727 in Rome in 1985, shot and killed a US sailor on board in Beirut and prompted President Reagan to send an aircraft carrier against them. The next year US bombers struck at Libya, then in 1988, five months after the *Vincennes* incident, Libyan agents struck back, not with their own military aircraft but by putting a bomb on board Pan Am Flight 103, which blew up over Scotland and killed 258 passengers and crew, plus a dozen in the town of Lockerbie below.

Aviation was now a huge, dog-eat-dog industry of the highest technological and organisational order. Even airliners, such as the Airbus A320 in 1987, started to become flown by computer. Airports were vast industrial complexes creating their own local economies. Air traffic control was now so programmed and complex that flights were an automated process, rather than an exercise of individual skill on the part of pilot and navigator. Only in the smallest planes and remotest locations could flying truly be called free and fun.

One 19-year-old in a small plane, however, did achieve something beyond the dreams of air marshals. In 1987, German pilot Mathias Rust flew from Helsinki under Soviet radar to land right in Red Square in Moscow, easily defeating what was supposed to be one of the most sophisticated air defence systems in the world. The flight turned out to be just another indication that the great Soviet edifice was cracking. By 1990 the Cold War was clearly over. What then, to do with the vast Western aviation arsenal?

The answer came when Saddam Hussein invaded Kuwait in August 1990 and so gave an excuse for almost the entire NATO air power doctrine to be put into effect. Everything in the arsenal was used in the 1991 Gulf War. And newest of all were the first 'invisible' planes, F117A Stealth bombers which used composite materials and sharp, flat angles in their design to deflect radar, dropped precision-guided bombs against high-value Iraqi targets. For the first time in history the air power beliefs of Douhet and Mitchell, Harris and LeMay seemed to be vindicated. Yes, troops were still needed, but arguably this was the first time that aviation led the operation rather than just supported it.

1993-2002

If the Gulf War proved anything it was the total dominance of US air power, and over the last ten years that has been reinforced. Bombing attacks on Serb targets in Bosnia greatly helped enforce compliance with the Dayton Peace Accords in 1995, and in the 1999 Operation Allied Force against Serbia precision guided bombing showed the tremendous leap air power had made since the Gulf War. One Tornado GR1 equipped with Paveway II bombs was able to do damage far greater than an entire squadron of Lancaster bombers in World War II.

Bombing accuracy was now so good that when things did go wrong it looked all the worse: the Al Firdos bunker in Baghdad in which many Iraqi civilians were killed, the F-16 attack on a refugee convoy in Kosovo and the attack on the Chinese embassy in Belgrade all became significant political issues. Aircrews increasingly realised that their tactical strikes could have strategic, highly political consequences.

In its own perverse way, the terrorist attack on America on 11 September 2001 brought the power of aviation back to its roots in Orville and Wilbur Wright's bicycle garage in Dayton, Ohio. The military call the attack an example of 'asymmetric warfare', whereby one side declines to fight in conventional ways, knowing it is unlikely to win, but instead attacks where it believes its enemy to be most vulnerable and with methods least expected. 11 September revealed that for all its immense sophistication, aviation could still respond to the will of the individual. To be sure, the terrorists had some pilot training, and it took them up to five years to plan and achieve their mission, but they proved that they could do damage to their avowed enemy and the world's economy out of all proportion to the effort and resources they put in. In a way, though, this was a highly symmetrical piece of warfare, because hijacking four civilian airliners over a peaceful United States was a perfect symbolic response to overwhelming American air power.

With the collapse of the Taliban government as a result of the 2001 US-led air campaign in Afghanistan it is being argued that air power alone is able to achieve results which, until recently, required thousands of troops on the ground. Air power is now undoubtedly the politician's first weapon of choice: relatively quick and casualty-free, and not entangling his nation with another in the way that ground troops do.

The overwhelming superiority of Western air power begins to raise questions: is the next-generation F-22 fighter really needed, for instance, or the British-German-Italian-Spanish Eurofighter – originally designed in the

Cold War – or a number of other major aircraft projects? Afghanistan saw the first operational use of an unmanned combat aerial vehicle (UAV), a US Predator controlled from the ground that launched a missile at a ground target. Is the era of the manned combat aircraft doomed? With the pilot's physiological limits probably already reached, the future is more likely to be in missiles and UAVs, with manned 'mother ships' flying some distance from the danger areas. The war-winning concepts today seem to have less to do with the aviation 'platforms' than with their associated technologies, the key being C4ISTAR (Command, Control, Communications, Computers, Intelligence, Surveillance, Aerial Reconnaissance). In industrial and political terms the high-tech aviation future is increasingly co-operative and international. Not even America wants to have such dominance that its allies cannot keep up.

In civilian airliners, too, global competition appears to be coming down to two large conglomerates – Boeing versus Airbus. Once again we are back to Europe's Concorde versus the American 747 – but this time in reverse. The question is whether the public will prefer, and economics favour, flying from hub to hub in the super-sized 600-seat Airbus A380, or from regional centres in smaller, but faster Boeing Sonic Cruisers.

After its first one hundred years aviation has reached a great many aspects of life, though it has become complex and expensive in the process. But for individuals there are still hot air balloons (even if one does not dress as elegantly as at Hurlingham in 1908, opposite), gliders, microlights and paragliders to re-create the simple freedom of being in the air that first attracted the pioneers one hundred short years ago.

8895.
L.S.&

High Flight

Oh, I have slipped the surly bonds of earth
And danced the skies on laughter-silvered wings;
Sunward I've climbed and joined the tumbling mirth
Of sun-split clouds – and done a hundred things
You have not dreamed of – wheeled and soared and swung
High in the sunlit silence. Hov'ring there,
I've chased the shouting wind along, and flung
My eager craft through footless halls of air.
Up, up the long delirious, burning blue
I've topped the wind-swept heights with easy grace
Where never lark, or even eagle flew;
And while with silent, lifting mind I've trod,
The high untrespassed sanctity of space,
Put out my hand and touched the face of God

John Gillespie Magee, Jr.

The 'aviator's poem' was written by 19-year-old Royal Canadian Air Force pilot John Gillespie Magee Jr after he flew a new Spitfire fighter at 30,000 feet over England in September, 1941. He was killed in a flying accident in December.

1903-1913

1 The dream is real

First came the gliding tests (right), but when Orville and Wilbur Wright put a 12 hp engine and propellers on their No.3 glider and flew their first Wright Flyer from Kill Devil Hills on 17 December, 1903, (preceding page), few people had any idea what they had done, though the love affair of the camera with the aeroplane was established from the first moment. Two years later even the US government still had not understood that the Wrights had the world's first powered, controllable, heavier-than-air flying machine. Worried about patent rights and not the showy kind anyway, the Wrights retreated to their bicycle shop in Dayton, Ohio. Europeans thought flight was an extension of motoring, and their first flight in 1906 owed more to box kites than aeronautics. Prompted by big prizes, however, by 1908 aviation fever in Europe was catching on, and the Wrights came over to show how it should be done. A year later the biggest aviation event of the decade, for Britain at least, came with Louis Blériot's flight across the English Channel. For the first time the island nation was accessible – and vulnerable – from the air. Manned aeroplanes remained a doubtful proposition to many, however, and experiments were still going on into flapping-wing ornithopters, gyroplanes and the like, while Count von Zeppelin's huge, rigid airships perhaps seemed a safer bet for passenger travel. Zeppelin, who had launched his first airship in 1900, formed the world's first commercial airline company, Delag, in 1909. But by 1913 aeroplanes had become more powerful and sturdier.

Glider No 1

Orville Wright lands the Wright No.1 glider badly, his brother Wilbur racing up to help. Their tests of gliders starting in 1900 built on the work of the German Otto Lilienthal, killed four years earlier. They established the principle of flight control through 'wing warping', but had yet to invent an effective rudder.

The Wright Brothers

Orville Wright (top) and his brother Wilbur (right), photographed with their latest Flyer at Auvours, France, in July, 1908. The Wrights came from a devoutly religious family in Dayton, Ohio, whose first joint enterprise was a family newspaper using a printing press made by Orville, the elder of the brothers. They started building bicycles in 1892, and gliders by the turn of the century, spending months each winter in the desolate sand dunes of North Carolina.

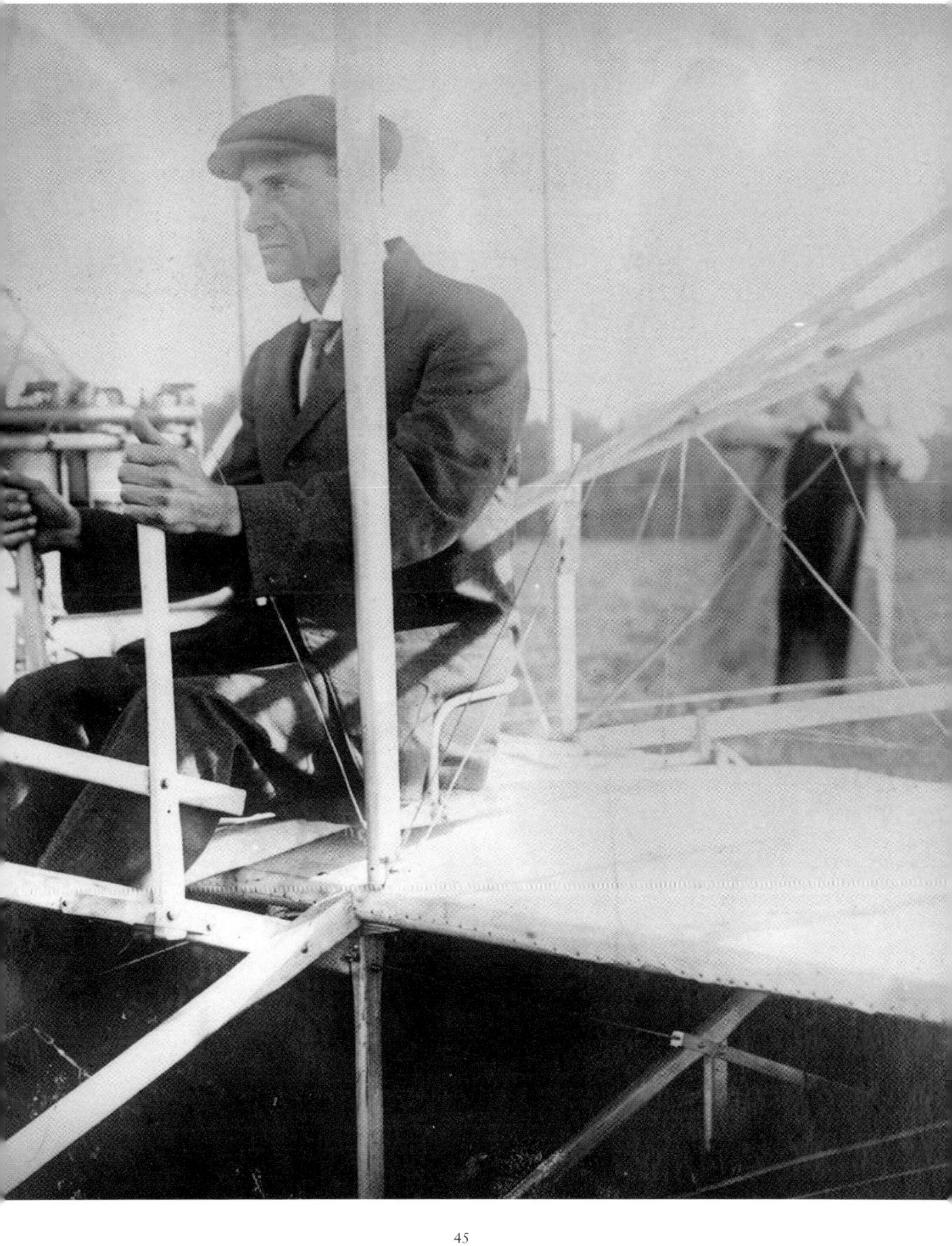

Europe's first flight
Brazilian aviator Albert Santos-Dumont, already famous for his balloon airship designs, stands up in his strange, tail-first box-kite biplane 14-bis for a 60-metre flight near Paris. The flight, in October, 1906, was Europe's first. Unlike the Wright Flyer, the 14-bis (so named because he tested it under his No 14 balloon) was designed more for stability than control. As with the Wrights in 1903, the cameras are there to record the moment of lift-off.

Box-kite pioneers

In January, 1908, Henri Farman (above, left) flew a record-breaking 1 km circular course in France in a box-kite biplane built by Gabriel Voisin (above, right). Nine months later people stopped and stared as Farman, again in a Voisin box-kite (left) made Europe's first cross-country flight, the 16 miles from Buoy, near Chalon, to Reims.

Marquis d'Ecquevilly's Multiplane, 1908

Guillon's Guillon and Clouzy machine, 1907

Jacques de Lesseps' Le Fregate, 1910

Robart's Papillon, 1909

Bertrand's Unic No1 R.B., 1910

There is no evidence that any of these wonderfully imaginative machines actually took off, though Monsieur Guillon did get the front wheel of his off the ground on Epsom Downs. Bertin's autogyro had a small top-mounted rotor blade, while the two wing-flapping ornithopters tried to emulate birds. Neither tilted-up nor tilted-down wings helped Comte Jacques de Lesseps or Monsieur Robart. Horatio Phillips's four-frame version of the single runner bean frame seen here, derived from a design dating back to 1884 did, however, fly for 500 feet in 1907.

Horatio Phillips' Multiplane, 1911

Collomb's ornithopter, 1908

Bertin's autogyro, 1908

Bartlett's Flapping Machine, 1911

Stoeckel's monoplane, 1909

Strong and light

Weight and power were the Holy Grails of the early aviators – less of one and more of the other. Leon Levavasseur's excellent Antoinette engines (named after the daughter of his industrialist sponsor, Jules Gastambide), powered his Gastambide-Mengin I in December, 1907, forerunner of the famous Antoinette monoplane (above). The wings were strong enough for Gastambide and Mengin to stand on. A 20hp two-cylinder Antoinette engine powered Alberto Santos-Dumont's tiny Demoiselle monoplane (left). From 1908 the Seguin brothers' Gnome rotary engine, though much lighter, provided much more power.

First helicopter
Hard to imagine quite how, but Paul Cornu's twin-rotor machine, with him on board, rose vertically for a few inches at Lisieux, France, and hovered there for 20 seconds on November 13, 1908. Because it did so unassisted, it is reckoned to be the first helicopter. A machine the Breguet brothers flew two months earlier was supported by four men at each corner. It would be another 28 years before the problems of helicopter stability were resolved.

Crossing the Channel

The 22 miles of sea between France and England must have seemed immense to Louis Blériot when he nursed his Blériot XI monoplane across the coast near Calais just after dawn on July 25, 1909. Odds-on favourite Hubert Latham had already crashed in the Channel and Blériot himself was still nursing a painful foot, burned in a flight a few days earlier (above). He flew at an average speed of 46 mph and the crossing took 40 minutes.

Victorious Blériot

Louis Blériot poses with his wife and plane at Dover Castle a day after becoming the first person to fly an aircraft across the Channel, July 25, 1909. The weakest entry in a three-plane race, he almost missed the English coast entirely on his way over, having left behind the French destroyer guiding him. One of his unsuccessful competitors, Count Charles de Lambert, (standing behind the wheel in an overcoat), made the first flight over Paris three months later. (The original glass plate negative is damaged.)

A horse, a train, a tower, a plane

A symbolic juxtaposition as Count Charles de Lambert flies his Wright biplane above the Eiffel Tower in October, 1909. De Lambert, who had been beaten across the Channel by Louis Blériot, won a prize for flying a circular route from Juvisy to Paris and back.

The winner

It was the era of the magnificent men and their flying machines. In 1911, French Lieutenant Jean Conneau (racing as 'Andre Beaumont') won cheers from the crowd at Brooklands, near London, as he came first in the five-day, £10,000 Daily Mail Round Britain Air Race (above).

Reims, 1909

Swooping low around a race-track marker, Eugène Lefebre demonstrates one of six Wright Flyers to an amazed public at Reims, France, in August, 1909 (above) The Wright brothers were absent, but the show demonstrated just how far aviation had come in such a short time.

Air show

Dapper British aviation pioneer Claude Grahame-White shows off his new 'Baby' model Voisin-Farman box-kite to a lady friend at the Reims air show in 1909. Although the Wright Flyers were aeronautically superior, the box-kites were more popular and regarded as safer because they were more stable. They had no roll control, however, and could only be flown in dead calm weather. The engine appears to be running because two men can be seen restraining the tail.

British beginnings

Texan showman 'Colonel' Samuel F. Cody (left) was one of Britain's most important aviation pioneers. Despite, or perhaps because of, his 'outsider' image he worked closely with the British Army and designed the nation's first successful aeroplane, Army Aeroplane No 1, in October, 1908. History now applauds him more for his courage and perseverance in stimulating 'airmindedness' in the British public than for his technical contribution to aviation. Cody was killed testing a new design in 1913. The first all-British aeroplane to fly with a British pilot was Edwin Alliott Verdon Roe's Avroplane at Lea Marshes, near London, on July 13, 1909. A few days later he flew it from Wembley Park (above). The machine was extraordinarily light, weighing 399 pounds, including Roe himself, had a tiny 9hp JAP motor cycle engine and its fuselage was covered in thick paper.

A tragic end

At the Bournemouth Air Week in July, 1910, Charles Stewart Rolls (above) put his Wright Flyer through its paces watched by a large crowd (above right). He tried to land at a spot close to the grandstand. At 70 feet in a crosswind he was too high and put the nose down sharply. The Flyer's elevator broke off and he crashed and broke his neck (below right). The first Briton to die in an air crash, Rolls was already a maker of fine cars (Rolls Royce), a co-founder of the Royal Aero Club , and the first to make a two-way crossing of the English Channel.

America regains the lead

In 1909 American Glenn Curtiss designed a float plane (above) which went on to be developed as a major seaplane for the US Navy. Curtiss (right) sits at the controls of a 1910 aircraft with prospective customer Lt John Towers. Curtiss' first design, the June Bug, was a compromise between the manoeuvrability of a Wright Flyer and the stability of a Voisin-Farman box-kite. He won the world speed record in its successor, Golden Flyer, at the first Reims international air show in August, 1909.

Flying from water

Pilot Jean Becu, flying Henri Fabre's remarkable hydroplane Canard, takes off near Marseilles in 1911. It had first flown a year earlier, in March, 1910. Fabre's design owed much to his friend Gabriel Voisin, the box-kite pioneer, who had been the first to take off from water in a boat-towed box-kite glider in 1905.

Carrier take-off

Lieutenant Eugène Ely takes off in his Curtiss biplane from a platform fitted to the cruiser USS *Pennsylvania* in January, 1911. He also landed on this ship, a much greater challenge. A year earlier, in November, 1910, he had been the first to fly from a ship to shore when he took off from the USS *Birmingham*.

Women pioneers

Women clamoured not to be left out of aviation. On July 8, 1908, a few days after a suffragette dropped leaflets from a balloon onto the House of Commons, Thérèse Peltier flew in Italy as the first woman aeroplane passenger, though here seen at the controls (right). 'Ordinary' women were discouraged from actually piloting a plane, so took up men's names, such as 'Spencer Kavanagh' (Miss Edith Maud Cook), the first British woman to fly, or door-opening titles such as France's 'Baroness Raymonde de LaRoche' (Elise de LaRoche). She was the first certificated woman in the world to receive a pilot's licence in March, 1910. New practices meant new fashions, as indicated by Hélène Dutrieu (top left) who wore an 'aviatrice's jupe' for flying comfort, and American pioneer Matilde Moisant (top right), in 1911 the second woman to gain a pilot's licence in the USA after Harriet Quimby.

Hendon shows

Hendon, near London, quickly became popular for air displays. At the 1911 show people found their bicycles useful to peep over the fence and avoid paying admission charges (above). They may have seen Claude Grahame-White's damaged flying machine being carted away by an older form of transport (far left). At the 1914 Hendon Air Pageant, with war only weeks away, the First Lord of the Admiralty, future Prime Minister Winston Churchill, with his wife Clementine, came to see the best of world aviation for himself (left). He was keen to expand the Royal Naval Air Service and was learning to fly, until his wife put a stop to that in May, 1914.

Racing at Hendon
Claude Grahame-White, winner of the first International Air Race at Hendon, passes over monoplanes waiting at a start line for their race. By 1913 Hendon had fixed facilities, including hangars, production areas and administration buildings.

Looping the Loop

As aircraft became more powerful robust pilots dared try ever-more dangerous stunts to please thrill-crazy crowds. Adolphe Pegoud won fame as the first to fly upside down in September, 1913. He repeated the trick at Brooklands, England, a few days later, his audience preferring to lie on their backs. Credit for the first loop should properly go to Russian Lieutenant Nikolaevich Nesterov.

Military interest

As aircraft grew in power and endurance the armed forces of all nations took increasing interest in them. France was first to form an Aviation Militaire in October, 1910. At Hendon in September, 1912, naval and military craft of various types were lined up for inspection. Two weeks later, Germany formed its Military Aviation Service.

Hanging on

By May, 1912, three years after officers on Salisbury Plain complained about aeroplanes 'upsetting their horses', Britain had a new organisation, the Royal Flying Corps, which had a military section and a naval section. In September, 1912, 16 of the RFC's aircraft, 12 from the military wing and 4 from the naval wing, took part in manoeuvres on Salisbury Plain. With little or no brakes, the planes sometimes needed human anchors being secured when the propeller was started.

Trainee pilots

This motley, but well-dressed group of young people standing in front of a plane at St. Petersburg in 1910 are cadets of the first Russian Association of Air Navigation. In this picture by famous Russian photographer Karl Bulla the first Russian woman pilot, L. Zereva, is standing third from the left..

First passenger plane

Making money from flying was obviously a key driving factor behind the growth of aviation. Aeroplane designers found room for more passengers, singly, and then in two's and three's, as engines grew more powerful and weight restrictions were eased. This 1912 aircraft, with the German pilot Siegfried Hoffman at the rear, is believed to be the first plane actually designed to take passengers.

All aboard

British aviation pioneer Claude Grahame-White prepares to take off in his five-passenger Aerobus at Hendon in 1913. By 1914 aviation aimed to compete with ships and trains to carry passengers faster and over longer distances.

C1114

1913-1923

2 War and exploration

With the exception of the British Vickers FB5 Gunbus, no aircraft were designed for combat by the eve of World War I in 1914. They were seen primarily as slow, stable platforms which could be used for reconnaissance. That quickly changed as bigger planes, originally intended to carry passengers, became bombers, and fighters – such as this US Army Nieuport 28 taking off during the Allied counter-offensive in 1918 (right) – were designed to shoot down planes and airships. The air war, vicious though it was, was nevertheless judged by the public to be infinitely more chivalrous than the slaughter in the trenches below, and forever established the image of the noble fighter pilot. The group photograph of the newly-independent RAF's No.1 Squadron (preceding page), with the fox terrier in a forage cap, an uncanny anticipation of Snoopy the Dog in his First World War leather flying helmet in Charles Schulz's *Peanuts* comic strip, and the scarves nonchalantly worn, is regularly duplicated by the modern No 1 Squadron posing with its Harrier jump-jets. The war enabled industrial production techniques to be brought to aviation. After 1918 bombers once again became airliners, and by 1923 great long-distance feats of flying were laying down what were to be the first global airline routes.

Aerial photography

The centre of the Belgian town of Ypres, site of three major battles during World War I, was already almost completely devastated in 1915 (above) by both bombs and shellfire. Aerial photography allowed detailed examination of the results of military action.

Reconnaissance

From the ground soldiers could see relatively little of the scope of enemy defences. Views of the Hindenburg Line, (above and left), however, from aircraft flying at 8,000 feet, revealed its shape, the German trenches and the effect of mines and artillery on it.

Evolution of the bomb

In 1914 aerial bombardment was a primitive affair, often involving a small bomb dropped by hand from a scout plane, such as this British de Havilland BE2 (above). By 1918 a score of incendiary or high-explosive bombs were routinely loaded onto multi-engined bombers. The largest bomb made and dropped by the British in the war was this 1,650-pounder (right), dropped by a Handley Page 0/400 on Le Cateau railway station on the night of 13 September, 1918.

9
8
7
6
5
4
3
2
1

Sad ends

On the ground millions of soldiers were slaughtered in World War I, but aircrew deaths were always regarded as something special, owing to the individual nature of air combat and the heroic stature of the early pioneers. In 1915, the dead British pilot (above), gazed on by German soldiers, would have had a highly respectful funeral. The German pilot of this plane (right) died after being shot down and crashing on a house roof behind his own lines.

French Nieuport 11

German Fokker DVII

British Sopwith Camel.

British SE5A

German Junkers J1

German Gotha GVb.

German Fokker E111

Germany's Fokker E111 Eindecker, a variant of the Fokker E1, introduced in mid-1915, was the first warplane fitted with a fully-synchronised machine gun, able to shoot directly between the rotating propeller blades. The Allies recovered from this 'Fokker scourge' with the help of such new planes as France's Nieuport II, which had a twin Lewis gun on the top wing. Germany's Junkers J1, a patrol plane, was the first all-metal monoplane in 1915, nicknamed the 'tin donkey'. From mid-1916 the German Gotha bomber began long-range attacks on French and then British cities. Britain's Sopwith Camel, named after the hump which covered the Vickers machine gun, was delivered in July, 1917 and is one of Britain's best-remembered fighters. The British SE5A was good too: a fast, tough, well-armed single-seat fighter and scourge of German squadrons. The Fokker DVII, however, inflicted huge damage to Britain's newly-independent RAF in April 1918.

The Red Baron

Baron Manfred von Richthofen, seen being helped from his Albatross biplane after a reconnaissance flight over the Western Front in 1916 (left), was probably the most famous flyer of World War I. A protégé of Oswald Boelke, and already with 16 Allied kills to his credit, Richthofen took over the elite Jasta 11 in January, 1916. Operating from French airfields such as this (right), Richthofen's 'Flying Circus' was dreaded but admired by Allied pilots. In 'Bloody April', 1917, the Royal Flying Corps lost 151 aircraft to the Germans' 70. Richthofen's red-painted Fokker triplane marked him out as a 'knight of the air' and when he was shot down and killed in April, 1918, having downed a world record 80 planes, he was buried with full military honours. His successor was Hermann Goering.

Roland Garros

Captain René Fonck

Liutenant Charles Nungesser

Albert Ball

Major Raoul Lufbery

Captain Georges Guynemer

Major James McCudden

Eddie Richenbacker

Roland Garros was France's first ace. He invented a gun which shot through an aircraft propeller, a secret lost to the Germans when he crashed behind enemy lines in 1915. He escaped, but was killed in 1918. René Fonck was France and the Allies' highest scoring ace, with 75 kills (three in one ten-second machine gun burst). He survived the war. Charles Nungesser of France was frequently wounded and shot down. He had 45 kills and survived the war. Albert Ball VC, holding the nose and propeller of his plane, shot down 44 enemy aircraft and was only 20 when he was killed in 1917. Raoul Lufbery was an American founder of the Lafayette Escadrille with the French in 1915 and had 17 kills. Georges Guynemer, perhaps France's most popular ace, had 54 kills and disappeared on a combat flight in 1917. James McCudden VC had 57 kills and died in a training accident in 1918. Eddie Richenbacker was America's top ace with 26 kills. He survived the war.

L23

Nulli Secundus

L23

R34

R34

From the turn of the century Count von Zeppelin had made his airships the pride of Germany. He launched his first airship, LZ1, in 1900, and in 1906 diplomats and members of the Reichstag came to Lake Constance for the launch of LZ3 (opposite, top). A later model is seen from shore (opposite below right). Britain's first military airship, Nulli Secundus, (opposite below left), did not last long, but her Zeppelin-inspired R-34 airship was the first to cross the Atlantic in July 1919. It arrived back at Pulham, Norfolk, (above), from Mineola, New York, after completing the first-ever two-way transatlantic crossing.

873
5

Zeppelins raided

The Royal Naval Air Service, assigned to defend Britain's coasts at the start of the war, didn't wait for Zeppelin attacks. On November 21, 1914, three RNAS Avro 504s took off from Belfort, France (left), and flew 125 miles up the Rhine valley to attack the huge Zeppelin sheds at Fredrickshafen (above). It was the world's first strategic bombing mission, and each plane dropped four 20-pound bombs, damaging an airship and its shed, and an adjacent hydrogen plant.

A civilian war

A cellar beneath a house in London is offered, for 'what shelter it affords', by its owner (right) in 1917. For the first time in almost a thousand years British civilians found themselves under direct, sustained attack from a foreign power. In Paris, whose history was less fortunate, an air raid device used in the trenches is tested on a rooftop (above).

THIS HOUSE
contains a
fairly good sized
CELLAR,
in the event of an
AIR RAID
passers by are
welcome to what
shelter it affords.

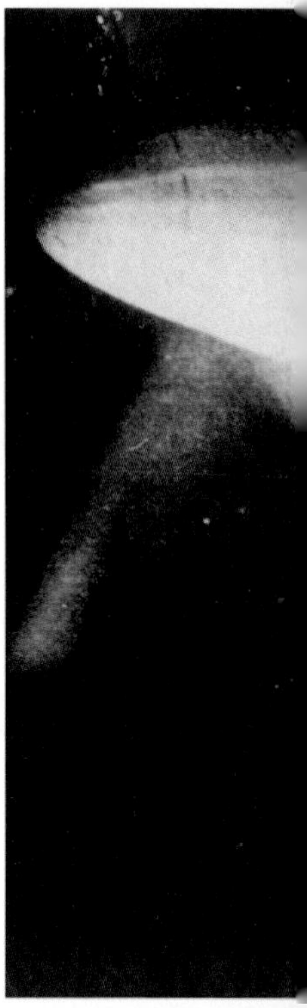

Bomb damage

Zeppelin raids caused most deaths and damage in Britain in World War I, as at Bartholomew Close, Central London, in September, 1915 (above). British defences improved, however, and Zeppelins caught in British searchlights (top right) became targets for Royal Naval Air Service fighters. Major Zeppelin raids on London ceased after 2 October 1916, when German navy airship L31 was shot down by a fighter at Potters Bar, north London, and plunged 12,000 feet in flames (below right). The first bombing of the capital by planes on June 1917 was also the worst single raid of the war. Fourteen Gothas dropped 72 bombs and killed 162 people, with 432 injured.

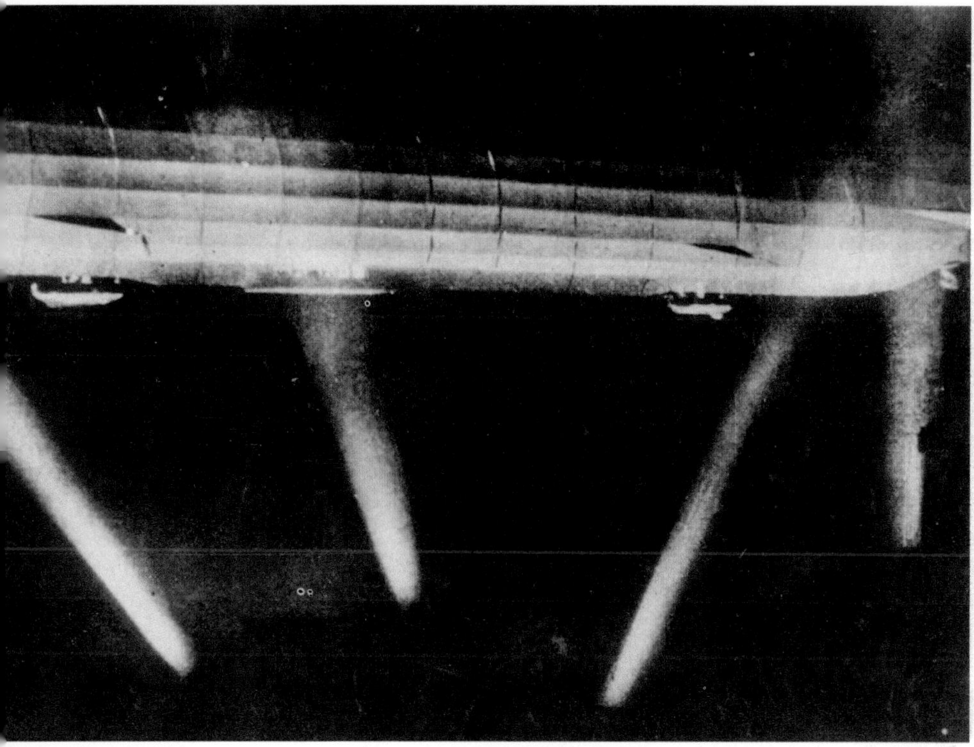

First carrier operations
Squadron Commander E.H. Dunning made the first landing on a ship under way on August 2, 1917, when he brought his Sopwith Pup down onto the deck of the first aircraft carrier, HMS *Furious* (above). He had a strong headwind and a little help from crew members. He was killed five days later trying to repeat his feat. HMS *Furious* demonstrated the strategic value of aircraft carriers when it approached the German coast in July, 1918, and launched seven Sopwith Camels (right) to destroy two Zeppelins in their sheds at Tondern, Germany.

Cold but not alone

Bomber aircrews were usually very cold on their long and uncomfortable bombing missions. The Handley Page 0/400 (right) had a maximum speed of only 97 mph and its four-man crew had to spend up to seven hours sitting in an open cockpit. Flyers quickly learned that formation flying (above) provided not only a concentrated force to strike targets, but allowed them to follow their leader and cover each other if attacked by fighters.

A 2

The watch on the Rhine
An RAF Handley Page 0/400 bomber of 48 Squadron flies along the Rhine and over the small town of Bonn, Germany in May, 1919, as part of Britain's post-war occupation duties. The war had ended just before new, bigger V/1500 bombers, capable of flying 2,500 miles or with a 7,500 pound payload of bombs, would have been used to bomb Berlin directly from England. Within a month this 0/400 was patrolling India's North-West Frontier.

Female industrial power

Women workers at Short Brothers seaplane works in England take a break (above) in January, 1919. War industries trained a huge number of women, and aviation brought them into areas of advanced technology which helped to change their status in the world.

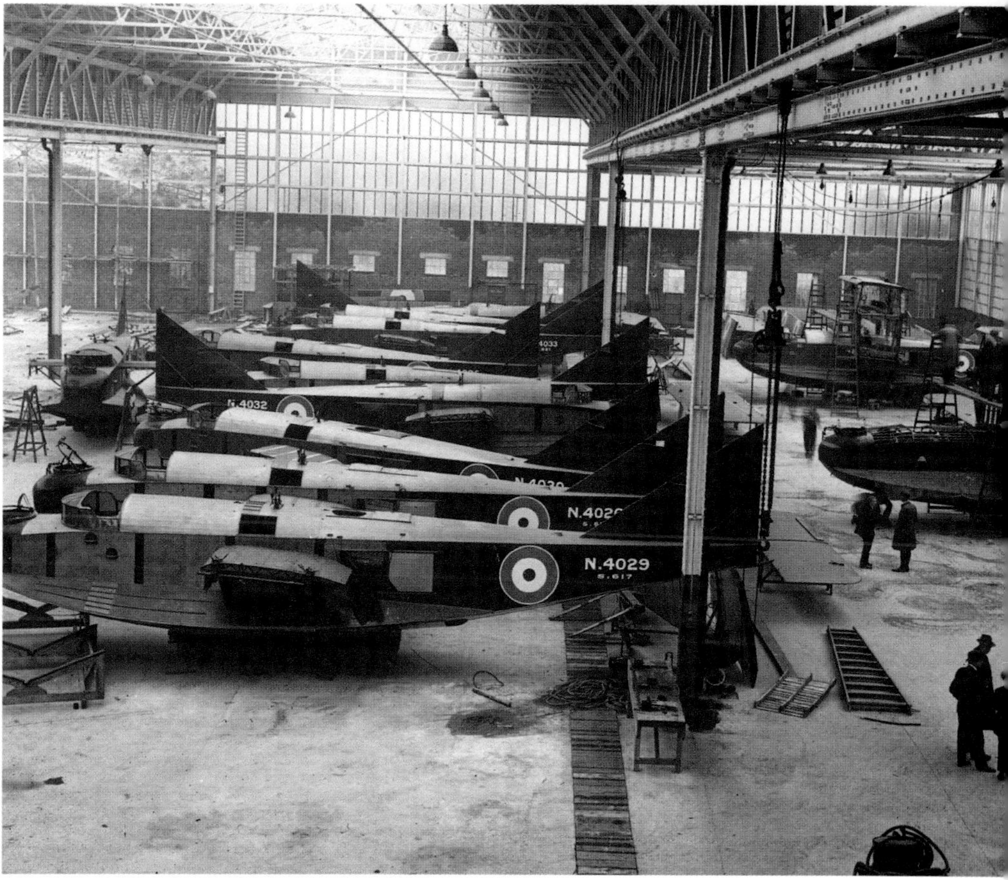

Aviation's industrial process

Until 1914 aeroplanes were crafted individually, largely by hand. Four years later aircraft were so advanced and sophisticated that aviation was beginning to challenge the shipyards in terms of skill and production techniques. Shorts' seaplane works (above) in 1919 illustrates the scale. The British aircraft industry that year employed 350,000 people. The RAF was the world's most powerful air force, with 22,000 aircraft and 300,000 personnel.

Transatlantic pioneers
A converted Vickers Vimy bomber sits damaged and partially-buried in a bog at Clifden, western Ireland, in June, 1919, at the end of the first non-stop air crossing of the Atlantic (right). British Captain John Alcock and Lieutenant Arthur Whitten Brown (above, at breakfast the next morning) completed the 1,890-mile flight from Newfoundland in 16 hours 28 minutes to win the £10,000 that had been on offer from the *Daily Mail* since 1913.

To Australia

In November and December, 1919, Australians Lieutenant Keith Smith, his brother Captain Ross Smith, and Sergeants W. Shiers and J Bennett (left to right) flew a Vickers Vimy 11,294 miles from London to Darwin in 28 days, winning a £10,000 prize from the Australian government.

French adventurers
Pioneer Sadi Lecointe departs from Monaco for the Persian Gulf in the early 1920s, attempting to set a new distance flight record (above), while Lucien Bossoutrot (right) arrives home in 1920 after being rescued with the six-man crew of his Farman Goliath after a forced landing on the coast of Mauretania. He had notched up a record 24 hours 19 minutes flying non-stop.

Pullman luxury

In an effort to compete with the London-to-Paris boat-train service, in December, 1919, Handley Page developed this Pullman airliner, complete with 14 seats and cushions, curtains and candle brackets (right). The London-Brussels route offered the first airline meals – lunch baskets at three shillings each.

AVRO

1923-1933

3 Coming of Age

The Roaring 20s was the decade that started to bring Flight to the Common Man. From spectator contests such as the Schneider Trophy and the US National Air Races to the Flying Doctor service in Australia and the first regular long-distance airline routes, an ever wider public started to see aircraft as the pioneers always intended – peaceful, quick, transportation. The glut of World War I flyers and aircraft continued well into the 20s offering ordinary people such as these holiday campers at Canvey Island, England, something extra with a joyride in an Avro 504 in 1923 (preceding page). At the luxury end this was the golden era for airships, especially the Graf Zeppelin. But in truth this was really America's decade, the era when the economic giant was stirred by one man in particular – Charles Lindbergh (right). His 1927 flight from New York to Paris in the 'Spirit of St Louis' electrified the world and galvanised the US into production and organisation of long-range aircraft that would lead it to dominate the world. Popular European design ideas, such as Anthony Fokker's high-wing monoplane, were snapped up and developed by US manufacturers. Indeed, the 1927 six-passenger Lockheed Vega, developed from a Fokker, became famous in the hands of such US pioneers as Wiley Post and Amelia Earhart.

Spirit

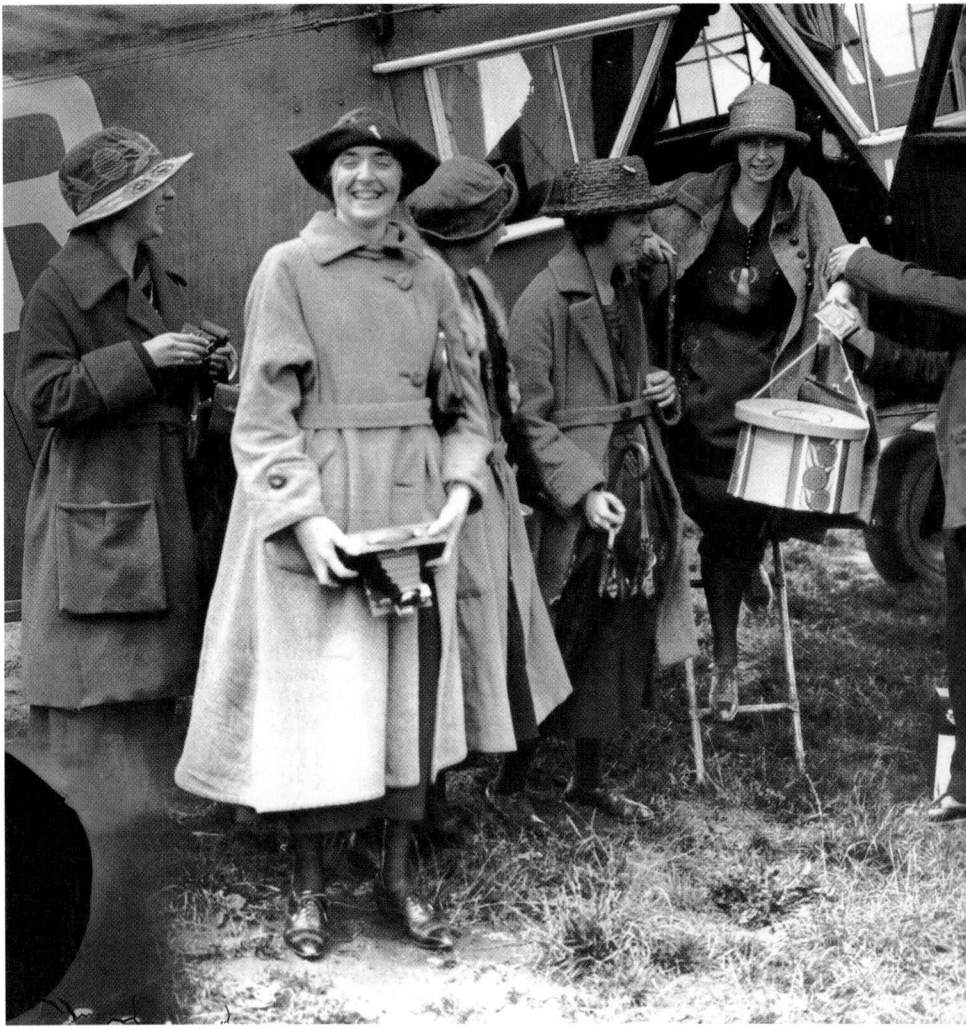

Wealthy passengers

Only the relatively rich, such as these American students arriving in London on an Instone Airlines flight from Paris in 1922, could afford to travel by air. Even so Instone had to have government subsidy to survive.

Golden cargo

A cargo of bullion from the Lena goldfields in Russia arrives at Croydon Airport near London via Imperial Airways in 1926, (above). High-value cargo such as this was a major reason why governments couldn't let airlines fail.

First in-flight movies

They were scratchy, single-reel ones but this group of passengers on a German airline in April, 1925, got to see the world's first in-flight movies. The films were, of course, silent but engine noise would have been too great to have heard anything anyway.

Highest German standards
German airlines were the envy of Europe in the 1920s. A number of first-class flights, such as this Lufthansa Berlin-Vienna Air Express in 1928 had cold meals, steward service and fresh flowers.

J.9406
9406

Air policing

The end of World War I and the collapse of the Ottoman Empire gave the British vast new areas of territory to police under League of Nations mandate. The newly-independent RAF eagerly sought to prove its worth at a time of drastic defence cutbacks, policing these areas without having to deploy divisions of troops. Westland Wapitis patrolled over the mountains of Kurdistan in 1934 (top), while the need to protect airfields and follow up sightings of trouble on the ground prompted the RAF to form its own troop unit, the RAF Regiment, seen with armoured cars on duty with a bomber in Mesopotamia (Iraq) in 1922 (below). RAF deployments in the Middle East between the wars involved five-year stints away from home, creating a closeness between air and ground personnel unsurpassed even in World War I.

East meets West

By the mid-1920s route pioneering was increasingly international, and by no means only from west to east. In September, 1925, Japanese airmen Abe and Kawachi were photographed in Berlin en route to London in the first Tokyo-London flight (above). A year later people flocked to visit a German Lufthansa Junkers G24 airliner after it landed at Harbin, China (top left) as it established the Berlin to Peking route. And in December, 1924, Dutch airmen with KLM airlines arrived in Dum Dum, outside Calcutta, India, en route from the Netherlands to Batavia, East Indies (left).

Distance pioneer

Britain's Alan Cobham earned a knighthood for his record-breaking and dangerous long-distance feats. In 1926 he made a 16,000-mile flight from London to Cape Town and back, establishing a route over the length of Africa (left). He then made a 26,703 mile round trip flight from London to Australia, returning to London on 1 October, 1926, and landing his DH-50 seaplane on the Thames opposite Parliament (above and right). He was, however, without his friend and engineer Arthur Elliott, who had been shot and killed by a Bedouin tribesman as they flew over Iraq.

G
G-EBFO

Transatlantic challenge

There were no longer any prizes for being the first across the Atlantic, but the $25,000 Orteig Prize for the first to fly non-stop from New York to Paris meant a journey twice as long as Alcock and Brown's.
Two of France's top war aces were involved. In September, 1926, the Mayor of New York, James Walker, christened Captain René Fonck's huge Sikorsky S-35 (above). Days later, however, it crashed on take-off at Roosevelt Field, New York (top right). Fonck survived, but two of the crew were killed. In May, 1927, French war hero Charles Nungesser and Francois Coli took off from Le Bourget, Paris, for New York in a specially-designed Levavasseur aeroplane (right) and were never seen again.

Lindbergh's arrival
Unassuming, handsome, ex-barnstormer Charles Lindbergh was a star the minute he landed his 'Spirit of St Louis' Ryan monoplane at Le Bourget airport, Paris, after a record-breaking flight from New York on 19 May, 1927. His fame was such that thousands waited for him at Croydon airport outside London on May 29 (right) when he was invited to lunch with King George V.

Air shows

Air shows were appreciated by the British public and seen by the RAF as a vital way for maintaining public support. The bombers flying low over the crowd at the Hendon air show near London 1926 in a display of 'evolutionary flying' were seen as part of the service's recruitment drive for its air and ground crews. The reason for such shows has not changed in almost 80 years.

Billy Mitchell

US Brigadier General William 'Billy' Mitchell (above), hero of the Saint Mihiel Salient in 1918 when he concentrated hundreds of Allied bombers to successfully assault a major German position, was enraged by cutbacks and the US Navy's insistence that aircraft did not pose a threat to warships. He demonstrated the launching of a torpedo by a Martin bomber in 1920 (right), and then the next year had the captured German battleship *Ostfriesland* hit and sunk by 2,000 pound bombs (top). It all failed to change navy minds in any fundamental sense. Court martialled and dismissed from the army in 1925, he was finally vindicated by the Japanese attack on Pearl Harbor in 1941.

A-5717
140

Naval air power

For all its refusal to accept the threat posed by strategic air power to itself the US Navy did understand the importance of aircraft to the Fleet. Its first aircraft carrier, the USS *Langley*, is seen with eight warplanes on deck during a show of naval strength off Baltimore in 1924 (top left). The battleship USS *Tennessee* tested out a catapult-launched floatplane from its deck during manoeuvres by the Pacific Fleet off California in 1926 (above.). The Royal Navy tested out a catapult-launched Parnall seaplane from a special hangar on the deck of the submarine *M2* in 1931 (left). The sub was to be lost, however, when the hangar door failed to close.

Parachutes
More than 6,000 British flyers were estimated to have died in World War I because they were not provided with parachutes, partly because it was thought they would not attempt to save their aircraft if allowed to jump. France felt similarly. Germany less so. By 1928 safe new parachutes were available and in that year all RAF aircrew were required to carry them. Training for a RAF pageant at Henlow, Buckinghamshire, in 1931 (above), and jumping from a light plane using a seat chute (right). These acted as a cushion upon which one sat in the cockpit.

Schneider Trophy

Hawker's Sopwith, 1919

Janello's Savoia, 1919

Long's Vickers Supermarine, 1931

Guazzetti's Macchi Fiat, 1927

The first post-war Schneider Trophy seaplane race in September, 1919, was dogged with controversy as Harry Hawker had to start from the beach at Bournemouth. Only the Italian Guido Janello was able to finish in his Savoia S13. Thousands watched at the Venice Lido in 1927 as Captain Fernando Guazzetti flew a Macchi Fiat M52 overhead. However, he retired with mechanical trouble. In 1931, Flt Lt F.W. Long was carried ashore from a Vickers Supermarine S6B, having won the 12th Schneider contest. Britain had also won in 1927 and 1929, so the trophy was now hers for good. The aircraft was designed by Reginald Mitchell, who used his Schneider experience to build the Spitfire.

Silent flight

The inter-war years saw a surge of public interest in gliders. A tiny Austin Seven car was used to tow a lady in a glider at Hanworth, England, in 1931 (top), while a 'pigeon' glider was captured on film after a heavy landing in a field near Paris in 1930 (centre). Herr Hentzen in his 'Vampyr' glider, constructed of wood and canvas, was a world champion in 1930 (below). He appears to be getting the maximum excitement out of the sport.

Gliding champs

The 'Scud' today is a long range missile, but in 1931 it was the name of a glider used in a British Gliding Association competition on the Sussex Downs (top). Long-distance gliding included a towed flight from Frankfurt by Miss Joan Meacon, seen being helped from the cockpit of her glider in Heston, London, in 1934.

Legendary women adventurers

Amelia Earhart, probably the world's most famous woman pilot (above), with her bi-plane 'Friendship' in 1928, was already a well-established air adventurer when she became the first woman to fly solo across the Atlantic in a Lockheed Vega in 1932. Her enduring legend began in 1937 when she disappeared without trace over the Pacific. Britain's Amy Johnson (right with her Gypsy Moth) was a more modest character who became a tremendous hit with the public following a record-breaking 19-day solo flight to Australia in 1930. A popular song, *Amy, Wonderful Amy*, was named after her. She died in 1941 when the RAF aircraft she was ferrying crashed into the Thames.

DH MOTH

Female power

Although flying and aviation were macho things, women were far from left out. Miss D Spicer cranked up the engine of her own plane at the first All Women's Flying Meeting in Northamptonshire, England, in 1931 (above). And while 'ladies who lunch' attracted a photographer's attention at a garden party for the London Aeroplane Club at Hatfield Aerodrome in 1935 (right) a dozen years earlier in America famous flying stunt acrobat Lillian Boyer (top right) was able to command fees of up to $3,000 a day – an enormous figure by today's standards.

LILLIAN BOYER

G ACNA

Aviation's new freedom

It was only open to the wealthy, of course, but if you had a light aeroplane of your own in the 1920s there was little better fun than to 'nip' across the Channel to France with your friends and set down in a field for a little picnic.

Competition flying

At Lympne, Kent, in October, 1924, light aeroplanes were lined up for the start of the Grosvenor Cup race. That year the British government helped to finance new flying clubs to encourage private flying. The long-term effect was to produce a pool of pilots who were desperately needed when war broke out in 1939.

The Graf Zeppelin
No airship was ever better than the *Graf Zeppelin*, the biggest at 775 feet. It made a huge show wherever it went, including the 1930 FA Cup Final at Wembley, London (left). The size of its gondola can be seen as boy scouts help with landing at Hanworth, near London, in 1932 (right, below). Inside, the *Graf Zeppelin* contained well-appointed cabins, kitchens and this luxury lounge (above right). After its maiden transatlantic flight to New York in October 1928, and a ticker-tape parade, its captain, Hugo Eckener, received a hero's welcome back in Germany. The *Graf Zeppelin* was the first airship to travel round the world. It flew until 1938, just after the *Hindenburg* disaster.

British airship luxury

The R-100 was Britain's attempt to compete in the long-range luxury airship business. The 709-foot airship, (left) seen at Cardington, Bedfordshire, in late 1929, was designed by inventor Barnes Wallis to a geodetic structure later used for Wellington bombers. Its hugely successful maiden voyage to Montreal in July, 1930, took 79 hours and carried 44 passengers in a luxury that included a grand dining room. (above). A privately-funded enterprise, it competed with the government-built airship, the ill-fated R-101.

End of the airship dream

For Britain, the end of airships as a serious competitor to long range passenger aircraft came at 1.07am on 5 October, 1930, when the R-101, on its maiden voyage to India, suddenly dived twice, the second time striking a low ridge at Beauvais, France. The engines were forced up into the gas bags and five million cubic feet of hydrogen exploded, leaving only the metal structure (right). Only six of the 54 people on board survived, not including the director of Britain's civil aviation. The sister ship R-100 was scrapped. The Germans persevered until 1937 when the *Hindenburg* Zeppelin exploded in flames on landing at Lakehurst, New Jersey (above).

Publicity shot

In 1932 the French Air Union's Golden Ray fleet inaugurated a new summer service from London to Le Touquet by bringing some London chorus girls to Croydon for a tabloid-style stunt photo. The girls came prepared in their beach pyjamas and swimming costumes to enjoy a swim at their destination. Note the pilot's open cockpit, even while the passengers were enclosed. It was still felt that aircrew needed exposure to the elements to appreciate wind and temperature changes.

The "GOLDEN RAY"

The TRANSCONT

1933-1943

4 Expansion into War

During this decade, half of peace, and half of war, the world of civil aviation was galvanised by the arrival of the Douglas DC-3 airliner (preceding page), and the military world by the Spitfire in the Battle of Britain. It was actually the low-wing, Junkers-inspired Boeing 247, which entered service in July, 1933, that was the first modern airliner. But then came the DC-1 and the bigger, longer-range DC-2, and it was Europe's turn to be shocked when the Dutch KLM airline dropped its own Fokker-designed planes for Douglas ones. Europe's primary choice in the mid-1930s, reflecting the increasing dominance of Germany's Lufthansa, was the tri-motor Junkers 52 and Ju 86. In Britain, aviation had never been so popular. One after the other local municipalities bought up flying club lands for city airports (right). Germany's blitzkrieg on Poland and the West in 1939 combined air and land forces working closely together in a way never seen before. France, cradle of European aviation, had dissipated its air power, and, despite valiant battles in which it destroyed 733 enemy aircraft, the Armée de l'Air simply could not compete. Britain's RAF held on with its superb Spitfires and Hurricanes, though RAF policy during the 1930s had left the navy with only a few old biplanes. But the weaknesses within the Luftwaffe's airfleets, caused by its hurried revival since 1933, were revealed. Naval air power was rammed down the throat of America by Japan in the attack on Pearl Harbor in December 1941, and then it was America's turn to hit back.

IMPERIAL

Going for comfort

Passengers on an Air France plane in 1935 had plenty of leg room and stylish service (above). As time went on the interiors of workhorse planes, such as the 13-seater all-metal Ford 5-AT Trimotor (opposite) were upgraded to modern standards. Affectionately known as the 'Tin Goose', this was launched in 1926 and became America's first successful transport aircraft. One flew scheduled routes to Lake Erie islands in Ohio until the mid-80s.

IMPERIAL AIRWAYS
LONDON

Speedy deliveries

Modern communications could hardly be quicker than those demonstrated in December, 1934, when a midget car ordered by telegram in London, was delivered in Paris just three hours later (opposite). Manufacturer H. Shillam rushed one of his 1hp cars to Croydon just in time for the 12.30pm flight to Paris. A few months earlier an Air France postal monoplane, the Arc en Ciel (above) was able to deliver British mail from London to South America in two days.

CHINA CLIPPER

Pacific Clipper

A Pan American Martin M-130 flying boat, the *China Clipper*, sits at its mooring at Pearl City, Hawaii, in 1938, ready to take up to 11 passengers on to China by way of Midway Island, Wake Island, Guam and Manila. Pan Am established its global reputation with its many flying boat Clipper services, including transatlantic. But they were not cheap. San Francisco to Manila in the Philippines cost $799 for the first 8,200-mile flight in 1936.

By sea

Long sea transits and a shortage of third world airfields encouraged development of flying boats. Germany's 12-engined Dornier DO-X, seen in New York on a round-the -world trip in 1931 (above), was one of the biggest flying boats. In October, 1929 it carried a record 150 passengers, a crew of ten and nine stowaways.

By land

Local armed guards position themselves near a British Imperial Airways Hannibal Class Handley Page HP42 airliner at a desert airfield at Port Sharjah, Persian Gulf, a staging point for air travellers to India in 1931. Only eight of the big, comfortable HP42s were built, including those of the Europe-service Heracles class, but over 16 years they carried 100,000 passengers without a fatality.

Flying boat luxury

A steward serves breakfast to passengers in sleeping accommodation on board Imperial Airways' 'Canopus' flying boat on the new Alexandria-Athens service in 1936 (right). Imperial bought 28 of the Shorts 'C'Class flying boats, each able to carry 17 passengers and two tons of mail at 164mph. Its interior, arranged with standard seating, is seen above.

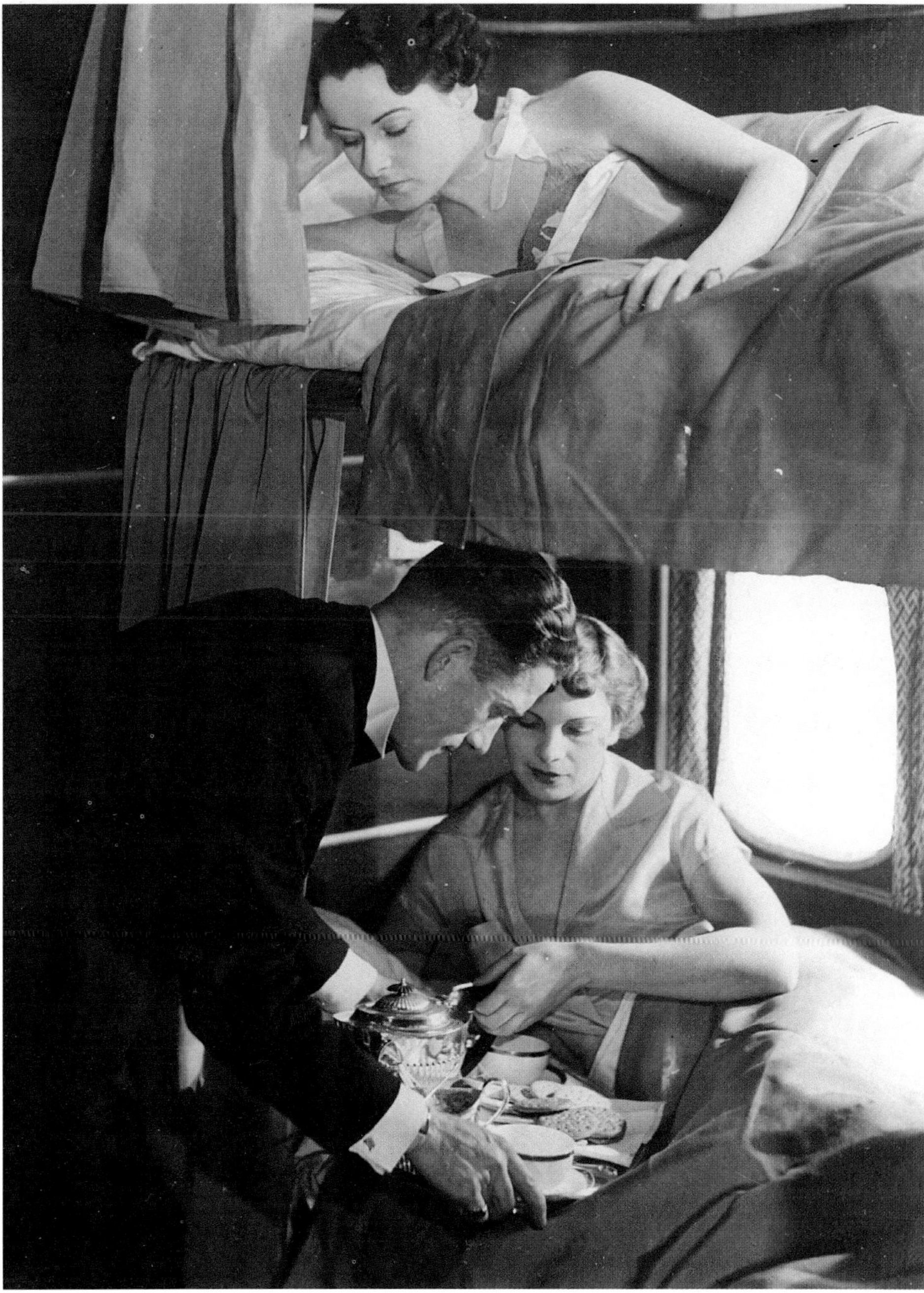

TRANSATLANTIQUE NORD LATÉCOÈRE 631

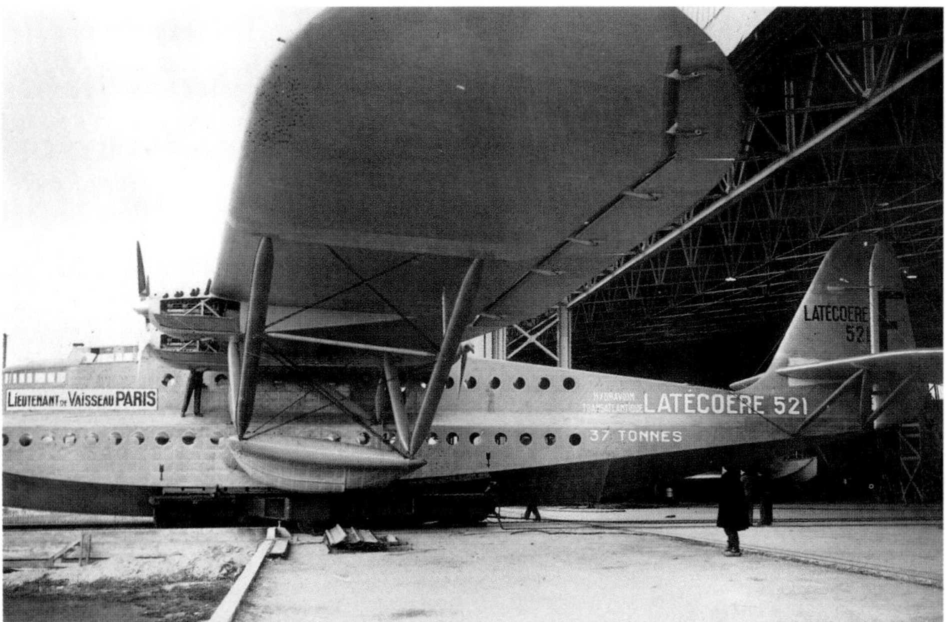
LIEUTENANT DE VAISSEAU PARIS
LATÉCOÈRE 521
37 TONNES
LATÉCOÈRE
521

Distance goals

Could a small plane beat a big one over distance? In 1934 British pioneer Sir Alan Cobham tested air-to-air refuelling over England (right, below) before setting off for a non-stop air-refuelled flight to India within 48 hours. In 1937 the 'Mayo' composite seaplane (top) tested a small seaplane carrying more fuel than would allow it to take off normally, launched from another seaplane in flight. Meanwhile, that year, the giant French aircraft *Lieutenant de Vaisseau Paris* established a distance record for flying boats for flying 3,612 miles across the South Atlantic (left below). The Germans confiscated this French Latécoère 631 46-passenger flying boat, but nine were built after the war (left, top).

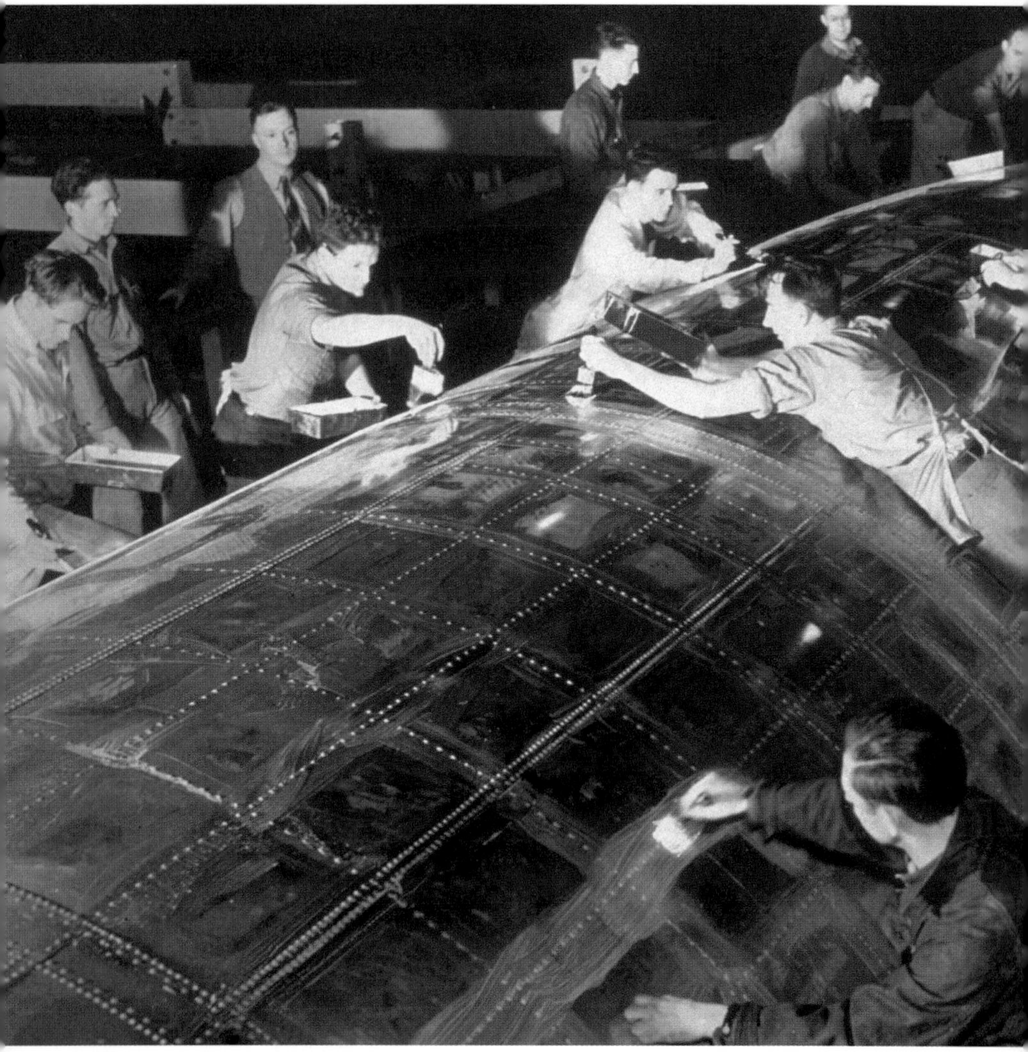

High-flyer

The 307 Stratoliner was the first pressurised-cabin passenger aircraft and could fly at 20,000 feet, so giving passengers a smooth ride above the weather. Workers at the Boeing plant in Seattle covered fuselage seams in soap (above) while the cabin was pressurised to ensure there were no leaks. If there was one, it showed up as a bubble. Only nine Stratoliners were delivered before war changed priorities for Boeing.

First modern airliner
The Boeing 247 (above) was the first's first modern airliner, introduced in 1933. Inspired by German designers Junkers and Rohrbach, the low-wing monoplane carried 10 passengers about 600 miles at a cruising speed of 160 mph. Boeing's next major innovation, with which the modern four-engine, long-haul aircraft saw its beginnings, was the 307 Stratoliner in 1938

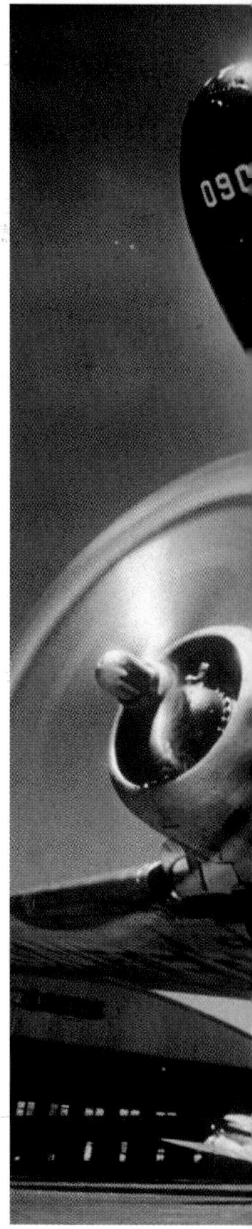

The fabulous DC-3

Probably the most famous airliner in aviation history, the Douglas DC-3 (seen opposite and above in the US in 1945) was introduced in December, 1935, and followed from the DC-1 in 1933 and the larger and faster DC-2 the next year. While the Boeing 247 was the prototype for the British Bristol Blenheim bomber and inspired both the German Junkers Ju 86 and Heinkel He 111 bombers, the DC-3 was faster, had wing flaps for better control, and carried 21 passengers 500 miles at 170 mph. Over the next 60 years some 10,655 DC-3s, also known as C-47s and Dakotas, were produced in the US for airlines and airforces around the world, with others built in Japan and under licence in the Soviet Union. Some are still flying in commercial or military use somewhere.

AIR LINES
CITY OF
PORTLAND

Airports

Jersey, Channel Islands

London, Croydon

Berlin, Tempelhof

London, Gatwick

Liverpool

The mid-30s saw an explosion of airports around the world as demand for air travel grew and flying boat bases were found to be too far removed from many large cities. Heavier aircraft meant expensive concrete runways had to be built to replace grass airstrips. Customs, ticketing and immigration facilities had to be established. In Britain municipalities took over many airport developments. Liverpool airport's terminal was built in 1938 with a control tower equipped with a radio guidance system to allow aircraft, it was claimed, to land in all weathers. Jersey's first airport was opened in 1937. Before then planes arriving at the island had to land on the beach. Gatwick airport, near London, was developed in the 1930s with a central 'beehive' terminal directly connected to a new railway station. Berlin's Tempelhof airport expanded while passengers arriving at Croydon from Dakar in a new French Dewoitine D.338 airliner in 1933 disembarked next to the terminal's well-known clock.

West Point of the air

Randolph Field, San Antonio, Texas, was where the US Army sent its pilots to study and train in the years leading to World War II. Here BT-7 trainers pull up in a patriotic salute.

Perfect line-up

US military pilots were, until 1946, in the army, so when it came to parading their training planes at Randolph Field they did so in strict soldierly fashion.

German testing ground

The Spanish city of Guernica entered the annals of infamy in April, 1937 when it was the first significant town to be smashed by German bombers in raids intended to terrorise and demoralise (above). Although terror bombing became more associated with siren-equipped dive-bombing Stukas, such as those shown in this German propaganda pamphlet about attacks on Crete in 1941 (opposite), Guernica was actually hit by wave after wave of Heinkel He 111 and Ju 52 bombers. Ju 87 Stukas were only operationally tested by the German Kondor Legion in the later stages of the Spanish Civil War. Pablo Picasso's painting *Guernica* has become the defining image of the terror of war from the air.

Stukas

Battle of Britain

Only RAF Fighter Command prevented Hitler from dominating the whole of Europe in 1940. Spitfires, such as this one seen from the rear gun turret of a German Heinkel He 111 bomber as it flashed past (above), joined greater numbers of Hurricane fighters to beat back hundreds of Luftwaffe aircraft in the Battle of Britain. The RAF was directed with the help of radar equipment around the coast which located approaching German air raids, such as these masts near Dover, seen in the background, during a German shelling attack on a British convoy (right). By the end of the war radar sets (above) were an established and vital part of any nation's defences.

Scramble

Pilots of the RAF's Free French squadron run to their Spitfires as German raiders approach England in January, 1942 (above). A Hurricane flashes past a Messerschmitt Me 110 it has just shot down over Goodwood, Sussex, in June, 1943 (opposite). Air and ground crews such as those of the Free French made Britain's air defences truly international. The Poles and Czechs were among the most dedicated and successful of Battle of Britain pilots, since many of them had already had combat experience in Poland and then France.

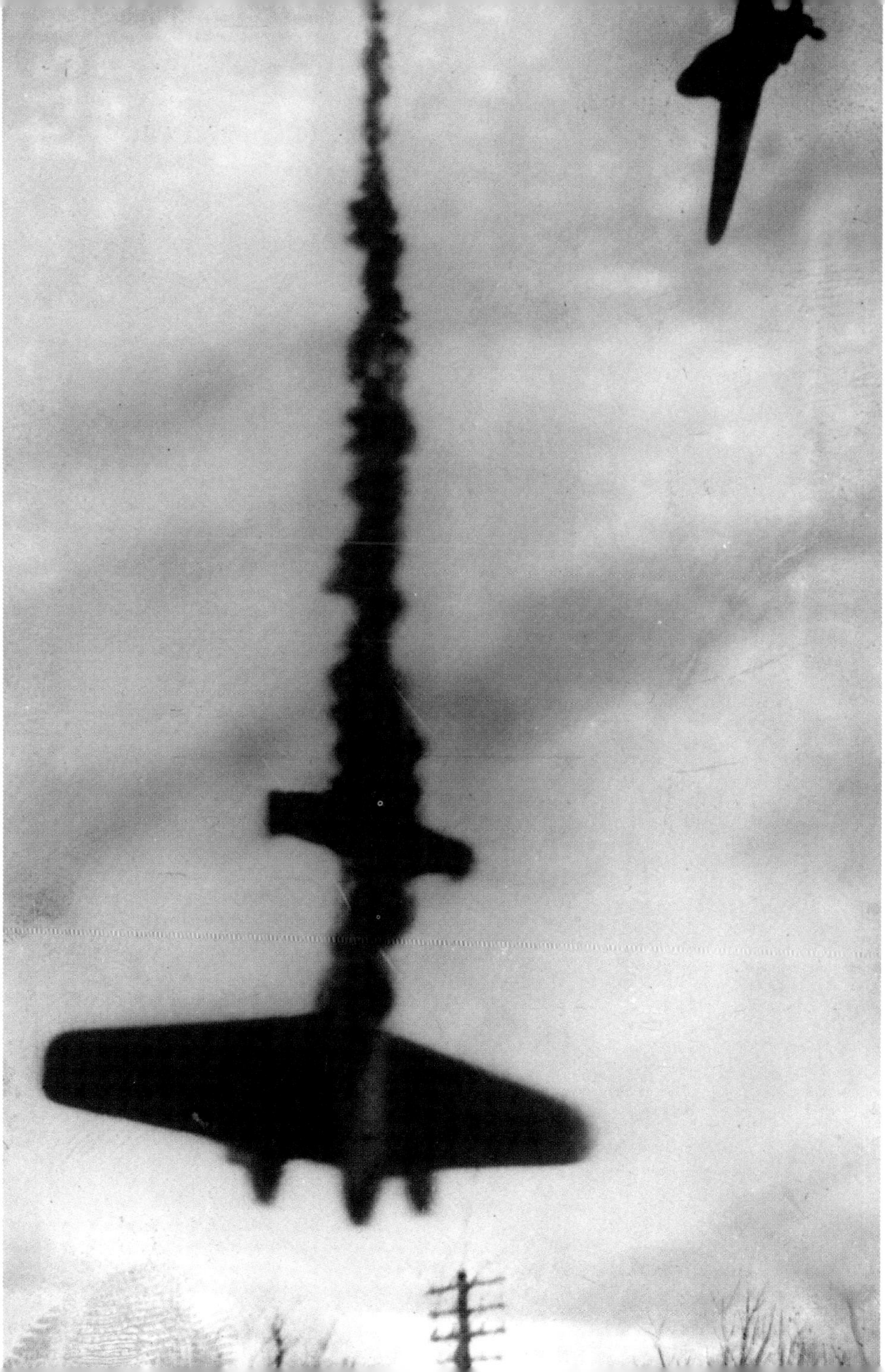

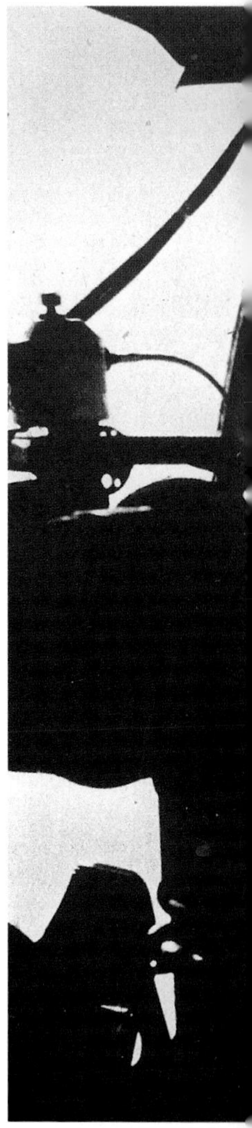

Dutch mission

A squadron of RAF Spitfires takes off from an English base for a sweep over Holland, led by Australian Wing Commander J.R. Ratten, in June, 1943.

A big sky
The view from the cockpit of a German Heinkel He 111 was undoubtedly exceptional, enabling the crew and bomb aimer to pick out targets with ease. The aircraft was slow and cumbersome, however, and the sight of a Spitfire attacking head on must have been terrifying.

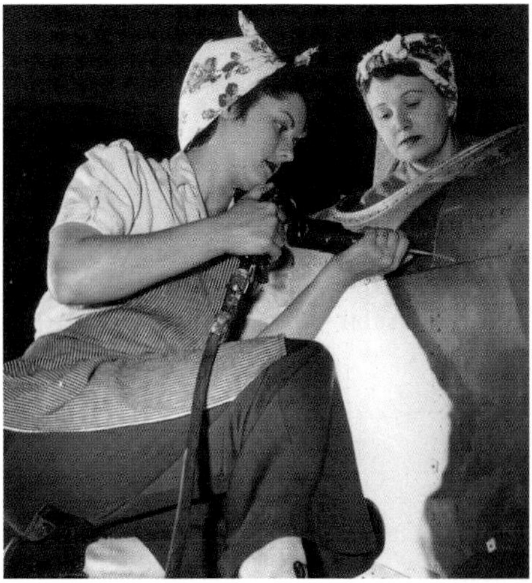

Women at war

First Officer Maureen Dunlop (opposite) ferried new aircraft as a member of Britain's Air Transport Auxiliary, The Women's Auxiliary Air Force plotted aircraft courses in the Operations Room at RAF Uxbridge, tightened cables attached to a barrage balloon, worked with an acetylene welder, while two factory workers used a riveter on a cockpit.

Eagle Squadron

Two Hurricane fighters from the RAF's new American 'Eagle' Volunteer Squadron fly over a colleague preparing to take off in March, 1941 (above). Although the US was not yet in the war, a number of Americans made their way to Britain to fight with the RAF. The Eagle Squadron survives today in the USAF, equipped with the F-15E.

The 109

Me (bf) 109s being prepared in Germany, March, 1940. The Spitfire had a tighter turning circle, better cockpit visibility and manoeuvrability, and its Merlin engine had slightly more power. The 109, however, was better in a dive and had better guns. Its biggest disadvantage was in not having enough fuel to stay for long over Britain. When it came to escorting RAF bombers over Germany, the Spitfire encountered the same problem.

Bomber's eye view
The view from a German bomber attacking London in the Blitz of 1940 is of two Dornier 217's flying over the Silvertown area of London's docklands. West Ham greyhound track is in the centre of the picture. The London Blitz, terrible though it was, may have saved Britain because it diverted the Luftwaffe's attention away from bombing fighter stations and gave the RAF the breather it needed to rebuild. The Luftwaffe was not trained and equipped to operate alone as a strategic air force anyway. Once the Battle of Britain was won, Hitler abandoned Operation Sea Lion, his invasion plan.

War heroes

Tobin

Rudenko

Hartmann

Bader

Douglas Bader lost both his legs in a flying accident in 1931 and was invalided out of the RAF. Against the odds, he argued his way back to flying in 1939, commanded a Canadian squadron during the Battle of Britain and shot down a number of planes. His spirit, including his truculence in German POW camps, earned him a lasting reputation. Captain Erich Hartmann, the Luftwaffe's leading ace, shot down 352 enemy aircraft, mostly obsolescent Soviet aircraft. Lieutenant Anatole Rudenko, a Russian ace. Pilot Officer Eugene 'Red' Tobin, in the cockpit of his RAF Hurricane was a founding member of the RAF's US-piloted Eagle Squadron in the Battle of Britain, but was killed over Boulogne in September, 1941.

Stringbag power

The war found the Royal Navy woefully short of modern aircraft, largely because until 1936 the RAF had control of the Fleet Air Arm and preferred not to spend its limited rearmament funds on naval air power. But obsolescent carrier-born Swordfish torpedo biplanes (right) showed their worth in November 1940, sinking half the Italian fleet in Taranto harbour. Then the German pocket battleship *Bismarck* caused a massive threat to British shipping in the North Atlantic in 1941, and sank the battleship *Hood*. At last light on 26 May and at extreme range a torpedo from a single Swordfish from the aircraft carrier *Ark Royal* (above) hit the rudder of *Bismarck* and caused it to go in circles. The Home Fleet caught up and destroyed it.

K
8428
M
K8428

The Eastern Front

The crew of a Soviet Tupolev Tu-2 medium bomber prepares for its next mission against German forces in Russia in the early 1940s. The Soviets did not design bombers for strategic military purposes and tended to use them as aerial artillery.

Stuka crews

German airmen with their Stuka dive-bombers just before the start of the Second World War. The Stuka was also used as flying artillery, diving onto its single target. It was the plane most associated with blitzkrieg, but was highly vulnerable to enemy fighters.

Tora, Tora, Tora

Japanese crewmen on board an aircraft carrier wave to the pilots of A6M and B5N fighter bombers as they set off for the surprise attack on Pearl Harbor. Some 365 fighters, bombers and torpedo aircraft took off from six Japanese aircraft carriers operating at a distance from Japan few thought possible. Japan took official British advice before and after World War I. Its first carrier was commissioned in 1922, and its first naval air strike was against Chinese airfields from the carrier *Kaga* in 1932.

Day of infamy

US sailors stand dazed amidst the wreckage of the Naval Air Station, Ford Island, Pearl Harbor under Japanese air attack on 7 December, 1941. Some 2,500 American soldiers, sailors and airmen were killed in what President Roosevelt declared a 'Day of Infamy'. The US also suffered the loss of five battleships, with three severely damaged; ten other ships sunk or damaged; and 188 of 394 military aircraft destroyed, with another 159 damaged. But none of the vital aircraft carriers was sunk.

Heading for Japan

Sixteen stripped-down B-25 Mitchell bombers took off from the carrier *Hornet* (above) on a one-way mission to Japan, dropping bombs on Tokyo, Yokohama, Kobe, Osaka and Nagoya. One landed in the Soviet Union, one in the sea and 14 in friendly China. The raid delighted America and outraged Japan, which ordered fighters back to protect the homeland. The planes were only able to take off thanks to the equivalent of a fifty-knot head-wind.

The Doolittle raid

Years after the US Navy rejected General Billy Mitchell's concepts for strategic bombers, it agreed in 1942 to place the aircraft carrier *Hornet* at the disposal of the US Army Air Force for a daring, morale-boosting raid on Japan. It was to be led by Lt Col James 'Jimmy' Doolittle, the US aviation pioneer who had won the Schneider Trophy for America in 1925 (above on *Hornet*).

The Coral Sea and Midway
Two US Navy SBD Dauntless dive-bombers from the carrier *Lexington* set off to search for a Japanese invasion force off the coast of New Guinea in May, 1942 (above). That month, in the first naval battle in which opposing ships never saw each other, aircraft battled it out over the Coral Sea and sank each other's ships 100 to 200 miles apart. The result was a draw: the *Lexington* was sunk but the Japanese turned back, having lost one escort carrier and with another severely damaged. A month later the two carrier fleets met again in the bloody battle of Midway. This time the Dauntless dive-bombers, despite heavy losses, took just five minutes to destroy half the Japanese carrier force. All 15 of the USS *Hornet*'s Devastator bombers were shot down, however, easy prey to the Japanese Zero fighters. From then on the Japanese Navy lost its superiority in the Pacific.

1943-1953

5 Momentous decade

This was the decade that saw air power impact brutally and brilliantly on humanity. It saw the development of strategic air power that made independent air forces equal or superior to armies and navies, the introduction of the first military and civilian jet aircraft, and the arrival of the first practical helicopters. By 1953 aviation was no longer a useful addition to a nation's development, but vital to it. But in 1943 that future seemed a long way off. Allied air power was struggling. Britain had poured its strained resources into RAF Bomber Command, sending up to 1,000 bombers a night on raids against German cities, raids whose value would later be questioned. America did the same by daylight and at equally high cost. But new, long range Mustang escort fighters started to turn the tide, and when Germany finally got its revolutionary Me 262 jet-powered fighter into action it was too late. The sacrifice of Japanese Kamikaze pilots (right) could not stop American air power, such as this American B-25 bomb run against a Japanese airfield in New Guinea (preceding page), and with the atom bomb dropped on Hiroshima in 1945 it entered a new dimension. Chuck Yeager broke through the sound barrier in 1947 and in 1952 Britain's Comet became the first jet airliner, while in 1948 a huge airlift to keep West Berlin fed defused the first major crisis of the Cold War.

War production

The roof lights of a Douglas aircraft factory in California reflect in the nose cones of scores of A-20 aircraft being produced in 1945 (right). B-17 cockpit sections are churned out en masse at Boeing's plant in Seattle (above). Allied aircraft production reached a peak of 167,654 in 1944; in spite of bombing, Axis aircraft production also achieved a peak that year of 67,987.

"MEMPHIS
BELLE"

Strategic bombing

US Eighth Air Force bombers, such as this B-17 flying over France in 1944 (above), flew daylight missions against German targets from England, while RAF Lancasters (left, top) led British raids at night. Few US bombers survived a full tour of 25 missions. One of the few was the B-17 'Memphis Belle' (left, below), with its tally of bomb missions and German planes shot down.

The Dambusters

In a hugely morale-boosting exception to the nightly grind of inaccurate bombing raids on Germany, on 16 May, 1943, Lancasters of 617 Squadron, a specially-formed unit, each carried a single cylindrical five-ton bomb to breach dams providing water to Ruhr factories. Against strong opposition several of the 19 Lancasters (above right) dropped the bombs designed by Barnes Wallis to bounce along the water and explode against the dams 40 feet down. The Mohne dam was breached (right), the water loss in the lake seen by RAF photo reconnaissance the next day (far right). The Eder and Sorpe dams were also hit. Fifty-three aircrew in eight aircraft were killed and the long-term impact on German war production was minimal. But raid leader Wing Commander Guy Gibson (above, centre) was awarded a Victoria Cross and by popular demand 617 Squadron remains a key RAF strike squadron to this day.

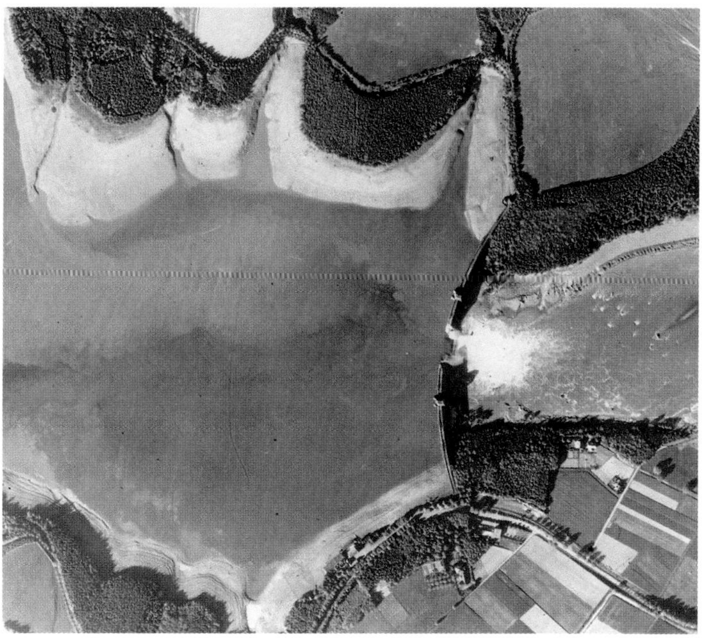

ANGELS WITH
DIRTY
FACES.
MAGARET
JOYCE
MARGIE
KATH
OLIVE
ELSIE

Airborne
Troop-carrying aircraft opened new possibilities of seizing positions before the enemy could react. The Germans used paratroops in the invasion of Crete in 1941 (right). The British and Americans subsequently formed airborne armies which played a big part in the Normandy invasion of June, 1944. British glider-borne troops give the thumbs-up before taking off on D-Day (left, top). The slogan on the glider refers to the dark camouflage paint they wear on their faces. In the British airborne attack on Arnhem, Holland, in September, however, (left, below) disaster struck when the lightly-armed Paras were confronted by unexpected German armour.

German terror weapons

Germany's V-I and V-2 rockets (V for *Vergeltungswaffen*, 'revenge weapons') opened a new era for aviation. The V-1 flying bomb (above) carried a 1,870 pound warhead and was first launched against London on 13 June, 1944. The 'doodlebug' had no target controls and its engine merely cut out after a flight time that would put it over the city. Civilian morale suffered as some 3,500 V-1's hit the London area, causing 6,184 deaths. The RAF's new Meteor jet was able to catch some V-1s thanks to their shallow trajectory, but not the more powerful V-2 rocket (left), which reached a height of 60 miles and speed of 3,600 mph. More than 1,000 smashed without any warning into England, killing about 2,700 in London. Another 900 were fired at Antwerp.

The jets

Gloster E28/39

Sir Frank Whittle

ME 262

Britain's Sir Frank Whittle had developed a revolutionary gas turbine jet engine, but it was German designer Ernst Heinkel, who built the first flying jet aircraft, the He 178 in August, 1939. His twin-engine He 280 became the world's first combat jet when it flew in April, 1941, coincidentally followed a month later by Whittle's first jet plane, the Gloster E28/39. His 600mph Gloster Meteor entered RAF service in July, 1944. But it was Willy Messerschmitt's Me 262 which proved to be the most dangerous jet. It entered service in June 1944 and shot down 427 Allied aircraft. It would have downed many more if Hitler hadn't insisted on trying to make it a bomber.

Ernst Heinkel

Kamikaze

By the spring of 1945 the Japanese had lost so many pilots that they turned to untrained or barely trained aviators and imbued in them a spirit of Kamikaze (Divine Wind) so they would fly suicide missions directly into attacking ships. One Kamikaze was caught on film a second before crashing into a US ship (above), while another was shot down in flames before he could make it (left). Admiral Chester Nimitz said that he was 'losing a ship and a half a day' to Kamikazes around Okinawa.

Nuclear end

US Col Paul Tibbetts (above) piloting the B-29 bomber *Enola Gay*, dropped the 'Little Boy' atomic bomb on Hiroshima on 6 August, 1945. A second atom bomb, 'Fat Man', was dropped on Nagasaki three days later.

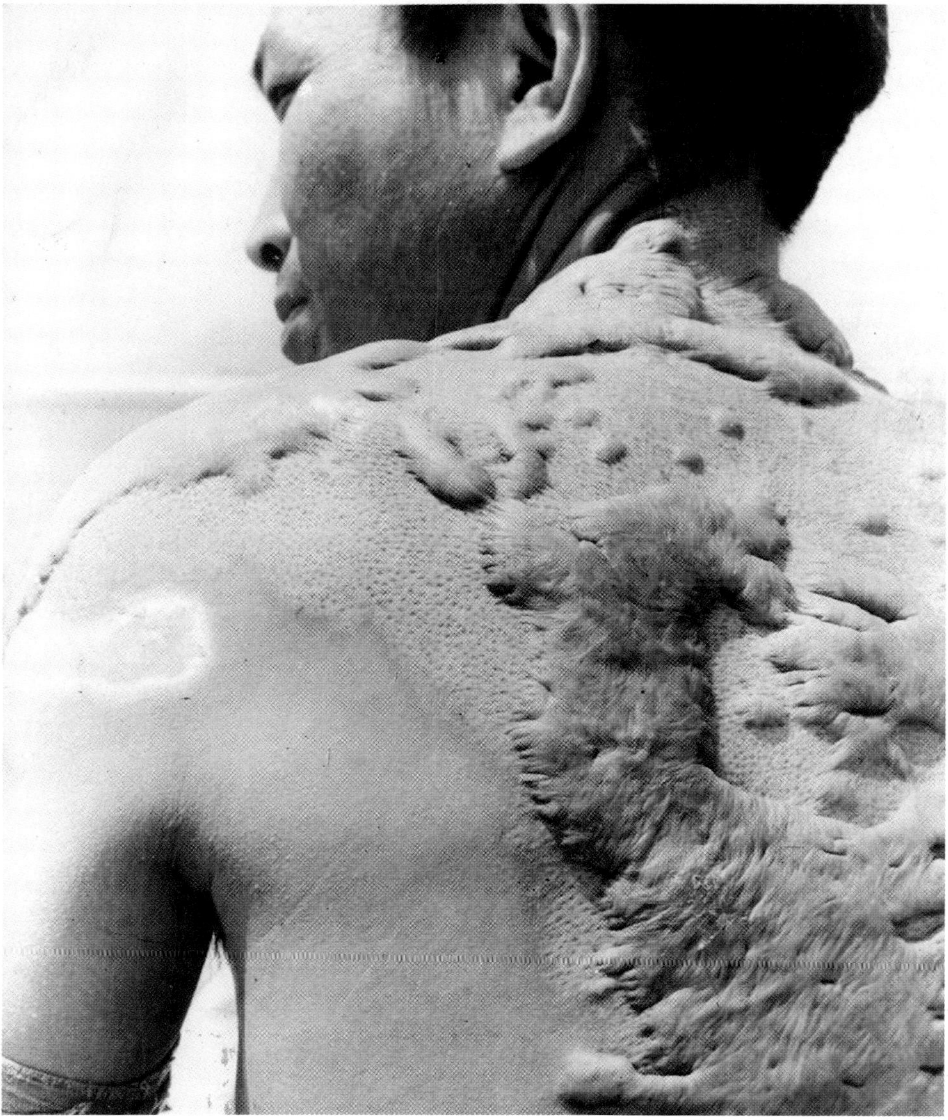

Radiation

A Japanese victim of the atomic bomb on Hiroshima reveals his radiation burns in 1947. It is not known how long he lived. Enola Gay's bomb killed about 78,000 people in the city, with many thousands more succumbing to radiation sickness later.

Hiroshima

Practically nothing in Hiroshima was left standing by the A-bomb. The bomb's power so affected British observer and war hero Group Captain Leonard Cheshire, VC, that he devoted most of the rest of his life to peaceful charity work.

'Conventional' damage

The ruins of Coventry Cathedral, England (above), are testimony to a concentrated night of bombing by 449 German planes in November, 1940. Sixty thousand of Coventry's estimated 75,000 buildings were destroyed or damaged and some 568 people were killed in the first evidence of a 'firestorm'.

Dresden

Women pass bricks for rebuilding Dresden, Germany, in 1946. A year earlier 773 British bombers, followed by 450 US bombers, had dropped over 3,000 tons of explosives on the city. Anywhere between 30,000 and 150,000 people were killed in the consequent firestorm, many of them unrecorded refugees.

DOUGLAS Flagship
WINGS FOR THE WORLD BY DOUGLAS
25¢ ADMISSION
Proceeds to Benefit
POLICE ATHLETIC LEAGUE
Recreation Program for Children of N.Y.C.
THIS AIRPLANE IS AWAITING ITS TURN TO BE MODIFIED FOR AIRLINE USE
Faisan d'Or
Airlines
FLAGSHIP NEW YORK
AMERICAN AIRLINES INC.

Douglas flagship

One hundred and thirty American Airlines stewardesses positioned themselves on and around its new Douglas DC-4 airliner, as part of a stunt, in June, 1946.The plane was hauled 15 miles from La Guardia Field to the centre of Manhattan. Douglas stopped making the 42-seat DC-4, originally designed in 1936, in 1948.

British giant

The Bristol Brabazon (above) was the largest aircraft ever built in Britain when it made its first flight from Filton, Bristol in September, 1949. Designed to carry 100 passengers across the Atlantic in comfort, the plane required Filton's runway to be extended to 8,250 feet, the detouring of a road and demolition of the village of Charlton in order to get it into the air. The dawn of the jet age killed it off.

Metal monster

Saunders Roe's Princess flying boat, seen outside a hangar on the Isle of Wight in 1951, started in May 1946. But the aircraft struggled to find favour and did not make its maiden flight until 1952. The 150-ton, ten-engine aircraft had 105 seats and was then the largest all-metal aircraft ever. Three were built, but they were not taken up.

Flying long range

While Pan American in 1949 introduced the 100-seat Stratocruiser, Lockheed already had a strong presence with its Constellation aircraft (left), which could carry about 50 passengers at 340mph across the USA and fly between New York and London in 11 hours. A group of British 'GI brides' and children arrive by Constellation in 1953 (above) for the Coronation.

British freighters
Three London taxis drive into the hold of a Bristol 170 freighter plane for a 20-minute flight across the Channel (above) in the late 1940s, while a Handley Page Miles 'Aerovan' is looked over at an air trade show at Radlett, England, in 1946 (left).

Viscount

Britain's initial response to the big US airliners was the Vickers Viscount (right), which introduced turboprop engines. These used a gas turbine, in place of a piston engine, to drive a propeller for vibration-free longer life, cheaper fuel and easier maintenance.

Chopper power

In February, 1948, a Sikorsky H-5 helicopter lowered food and mail to the Wolf Rock lighthouse off Land's End, England, cut off by a raging Atlantic for 26 days (left). In June another Sikorsky unloaded mail at King's Lynn, Norfolk, in a test of helicopter postal services (above). In August, 1953, Sabena, the Belgian airline, introduced the world's first international passenger helicopter service (right).

Berlin airlift

The Soviet Union's decision to squeeze the Allies out of Berlin by closing all road, rail and water access from West Germany in 1948 posed a tremendous challenge to supply 2.4 million people by air. By working night and day, aircraft such as this British DC-3 (above) and this US Air Force Douglas C54 Skymaster (right) kept the lifeline open. It took 277,569 flights to carry 2.3 million tons of cargo, and the achievement would only be surpassed by the airlift to surrounded Sarajevo, Bosnia, in the 1990s.

106
UR
60
NAVY
SL
NAVY

106
UR
60
NAVY
SL
NAVY

Korea rescue

Military helicopters came of age in the Korean War of 1950-53, as in this rescue of a US pilot who crashed off the carrier *Block Island* (left). Above all, Bell Sioux helicopters, with their stretchers fitted on either side of the bubble canopy, had their fame perpetuated by the *M*A*S*H* television series.

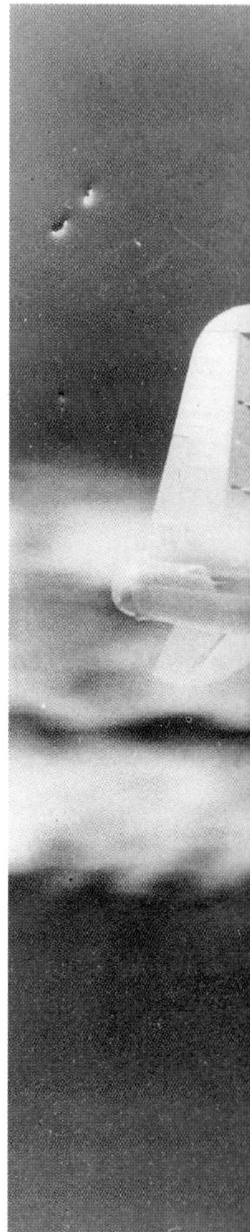

Going supersonic
World War II Mustang pilot Charles 'Chuck' Yeager (left) broke the sound barrier with a 670mph flight in an X-I rocket plane dropped from a B-29, mother ship' at 42,000 feet over California in October, 1947.

Faster still

In 1951 a Douglas Skyrocket launched from a Superfortress (above) hit 1,300 mph and in 1954 the US Air Force had the X-1A and X-1B (left, above) able to reach 1,650mph at an altitude of up to 90,000 feet.

Cold War build up

America's readiness to confront the Soviet Union is epitomised by this photograph of a B-47 Stratojet bomber of Strategic Air Command at March Air Force base in California in 1951 (right). Britain was caught short by the Cold War and had to use B-29 'Washington' bombers for a time. By 1952, however, these new Canberra jet bombers at RAF Binbrook, Lincolnshire, (above) had to come into service.

U.S. AIR
STRATEGIC AIR COMMAND

1953 – 1963

6 Era of the Big Jets

The big military jet bombers were here, but when BOAC's first de Havilland Comet was put into service in May 1952, it proved to be a false dawn for the jet airliner. After three fatal crashes in two years it was grounded. By the time it re-emerged as the bigger, better Comet 4 in 1958 Boeing was ready to beat it with the 707 airliner. US competition was too great for the private British plane-makers, and while the French government was supporting the monolithic Aerospatiale, by 1960 most of the great old British pioneering aviation names had merged into two big corporations, Hawker Siddeley and the British Aircraft Corporation. Passenger demand then was growing rapidly, particularly in America. Airport improvements were significant. TWA was the first to pamper passengers by building a terminal at New York's Idlewild airport with a moving walkway and covered ramps providing direct access from gate to aircraft (preceding page). In the military world the Cold War deepened, and huge nuclear-bomb laden B-52 bombers (right) of Strategic Air Command provided the main NATO deterrent to the Warsaw Pact. Britain joined the nuclear club and built a force of V-bombers. Then, in 1957, it decided that the future of military aviation was really in missiles and switched its aviation priorities. There was, indeed, a 'missile gap' between Russia and America around 1960 as each side built nuclear ballistic missiles, and when Gary Powers' high-flying U-2 spy plane was shot down over the Soviet Union by a surface-launched missile in 1960, the future of manned aircraft in high-intensity warfare did not look rosy.

U.S. AIR FORCE

First jetliner
Employees of the British Overseas Airways Corporation wave to the crew of a Comet 1 airliner as it leaves on the world's first jet passenger service, to Johannesburg, in May, 1952. It was powered by four de Havilland Ghost 50 turbojet engines and carried 38 passengers 1,750 miles at 490mph. A series of crashes, caused by structural weakness around the windows of the pressurised passenger cabin, took Comets out of service two years later. The improved Comet 4 replaced them in 1958.

B·O·A·C
G-ALYP

PAN AMERICAN
JET CLIPPER ENDEAVOR
PAN AMERICAN

America's first jetliner

Boeing's hugely-successful Model 707, the plane that dramatically changed the face of civil aviation, was America's first jet airliner when it entered service with Pan American in October, 1958, only a few days after the Comet 4 became the first commercial jet to fly the North Atlantic. It flew directly from New York to Paris (the Comet flew from London to New York and had to refuel in Newfoundland), carried twice as many passengers and flew 100mph faster. The 707 was first seen in 1954 as the 'Dash 80' prototype and quickly won massive orders from the US Air Force as the KC-135 Stratotanker before being sanctioned for a commercial version.

Jet competitors

In 1956 France's first jet airliner, the Sud-Aviation Caravelle (top) made its debut in Air France markings at Orly Airport and was christened '*Lorraine*' by Mme Charles de Gaulle. The innovative 80-seat rear-engine aircraft became a leader in the short-haul market. The Douglas DC-8 (above), introduced in 1959, was a significant rival to the Boeing 707.

Russian surprise

While the Comet 1 was being redesigned Russia amazed the West in 1956 when Andrei Tupolev's 50-seat Tu-104 jet airliner made its first appearance at London's Heathrow airport. Tupolev followed that with the 225-seat Tu-114 in 1959. The contra-rotating propellers, rarely seen in the West, were popular in large Soviet aircraft for aerodynamic reasons associated with powerful engines.

Noisy debut
Employees of Vickers aircraft at Weybridge, England, blocked their ears in June, 1962 as the company's first VC-10 airliner took off. The aircraft's four very powerful Rolls Royce Conway engines made a lot of noise on full power. The rear engines made the cabin quieter than in standard aircraft with engines under their wings, and also left the wings uncluttered for greater aerodynamic lift. The drawback was their relative inaccessibility high off the ground. The VC-10 later became both a tanker and a VIP aircraft with the RAF.

Light planes boom

Demand for light aircraft also boomed in the 1950s. US Cessna aircraft (above) became a household name around the world, the 1955 Model 172 in particular for 20 years becoming almost the standard four-seater in the western world.

Spreading the word

High-winged light planes, such as that owned by the Moody Bible Institute (above), took on the role of air taxis for more and more businesses and individuals who didn't want to have to use big airports.

USAF
USAF
USAF

Delta power
Four USAF Convair F-102A fighter interceptors make a striking image against the tyre marks of landing aircraft as they prepare to take off at an airfield in California in 1955.

42676
U.S. AIR FORCE

Extending the range
A Boeing KC-135 tanker refuels a Boeing B-52 Stratofortress high over the Pacific northwest in 1957 in a test to extend the range of the nuclear bomber. The KC-135, the USAF's main air refuelling tanker for the next 40-plus years, uses a boom probe controlled by a crewman at the rear of the plane who 'flies' it to the waiting aircraft. The French Air Force also uses the system, but the British and the US Navy use the 'probe and drogue' system which leaves it to the pilot of the tanking aircraft to locate and connect to fuel lines trailing from the wings of the tanker.

The Victor

Britain's Handley Page Victor bomber was one of a triad of nuclear bombers in the 1950s and 60s. The 1959 B-2 version (right) had a larger wing span and larger air intakes for its Rolls Royce Conway engines than the 1952 B1 version. The modifications were particularly intended to accommodate the Blue Steel standoff nuclear bomb (above), which could be launched some distance from targets and so was intended to improve the chances of getting a bomb through the strengthening defences of the Soviet Union. Victors served as air refuelling tankers in the 1991 Gulf War.

WZ364
WZ372
WZ377

V-power

The first of Britain's nuclear V-bomber force was the Vickers-Armstrong Valiant (left), a bomber and photo-reconnaissance plane. The second was the Avro Vulcan (above), which entered service in 1957 and was the last of the V-bombers to be retired in 1992. The Vulcan's huge, delta-wing shape and deep, powerful engines have made it one of the world's most pleasing air-show aircraft – particularly with women – and much effort has been made to restore the last one to flight.

Bothering a Bear...

The Soviet Union had its own long-range bombers, also used for reconnaissance, from the mid-1950s, and they remained in service through the 1990s. This Tu-95 'Bear' was intercepted over the Faroes by a US Marine Corps AV-8A Harrier in the 1970s. In 1972 two Tu-95s flew regular surveillance missions from Cuba along the east coast of the USA.

...and a Badger

The TU-16 'Badger', seen being intercepted by a US Navy F-4 Phantom over the Norwegian Sea in 1964 (above), was the origin of the TU104 jet airliner which shook up the West in 1956.

Family portrait

The US Air Force arranged this collection of representatives of all its major combat and support aircraft at the Air Proving Grounds in Florida in August, 1957.

U.S. AIR FORCE
U.S. AIR FORCE
U.S. AIR FORCE
U.S. AIR FORCE
USAF
U.S. AIR FORCE
U.S. AIR FORCE
5022
1210
10046
4270
21985

Vertical experiments

Flying Bedstead

Bell Moon Explorer

Hiller 'Flying Platform'

Flying Coleoptere

Using jets for vertical flight proved to be a major test of power and control. France experimented with the Flying Coleoptere in 1958, while a year earlier Britain's Shorts Co. tried with the five-engine SC1. In America, Convair's XFY-1 (an exception in that it was not actually jet-powered), made its first transition from vertical to horizontal flight and back again in November 1954. The US Navy tested out a Hiller 'Flying Platform' autogyro in the 1950s. While the British Fairey Jet Gyrodyne didn't catch on, and neither did the US two-man taxi designed by Bell Aerosystems for moon exploration in 1967, Britain's 'Flying Bedstead' in 1955 was a vital test rig for Rolls Royce vertical lift engines and shaped the design for the Harrier jump jet.

Shorts SC1

Convair CFY-1

Fairey Jet Gyrodyne

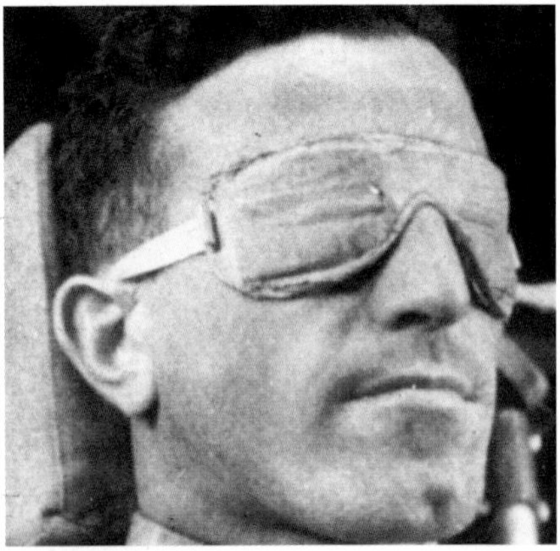
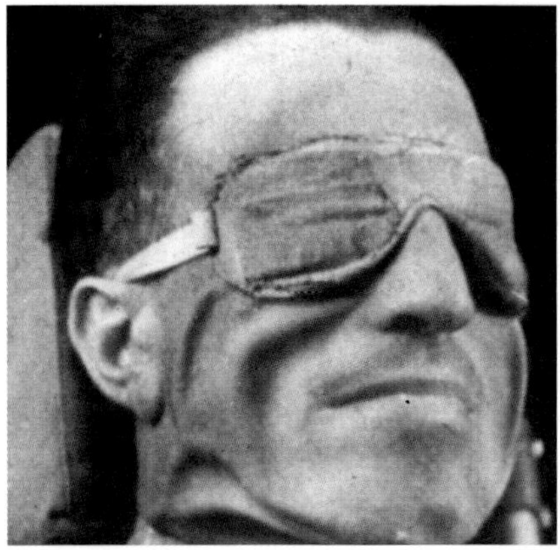
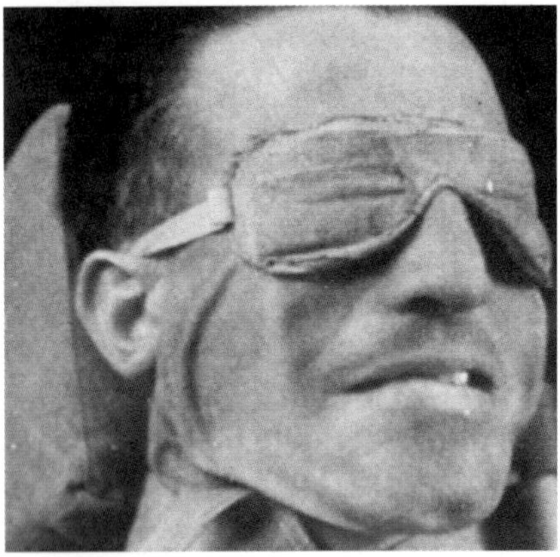
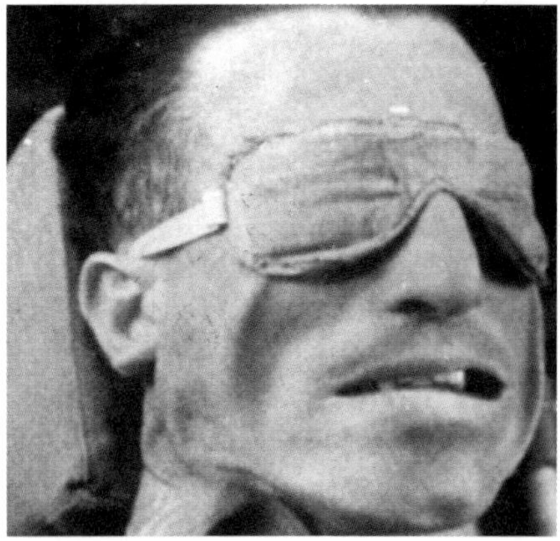
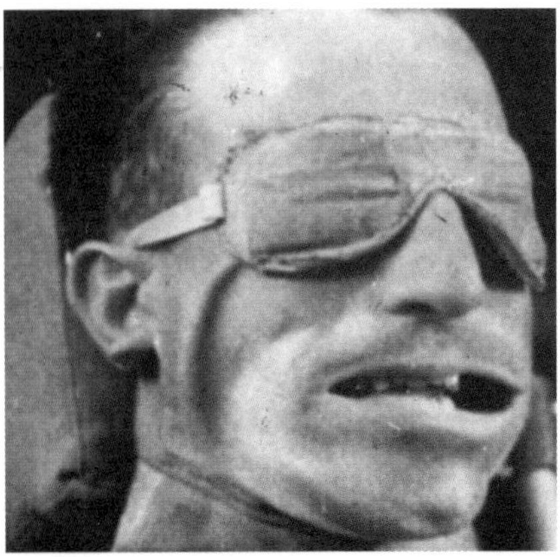
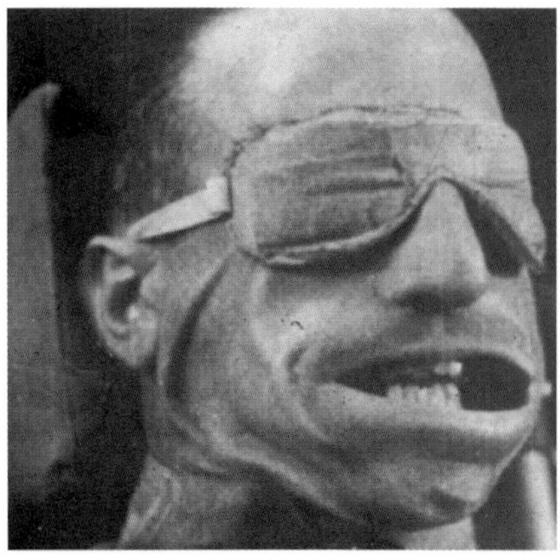

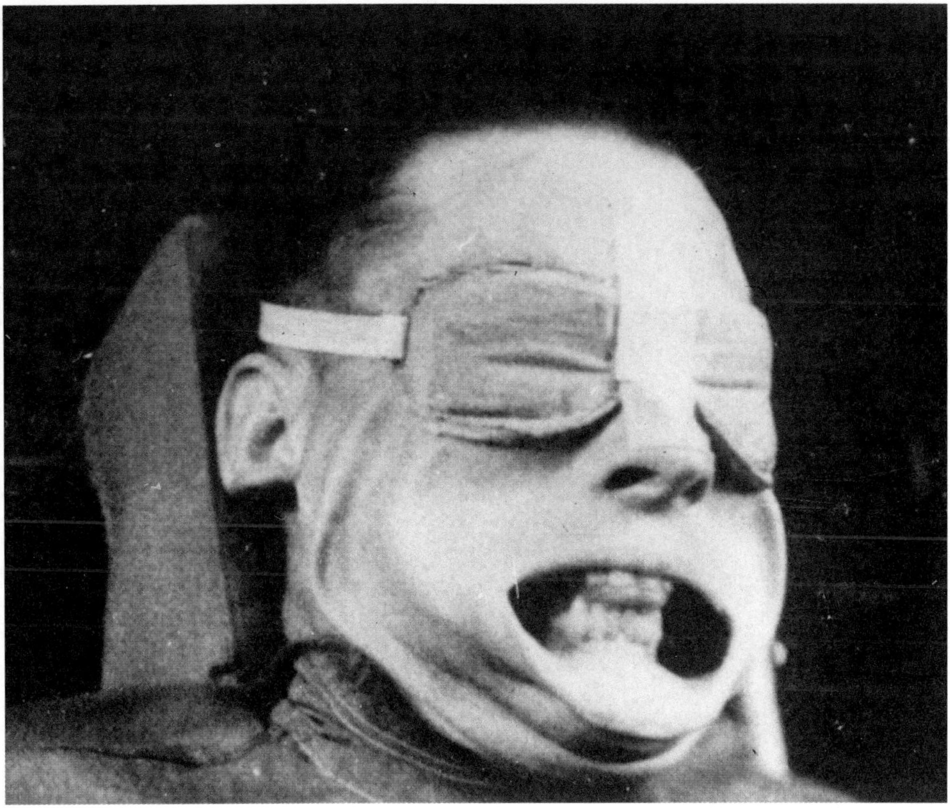

Wind blasted

The impact on a human face of wind blasting at up to 310 mph is shown in this progressive series of photographs at the US Naval Air Materiel Center, Philadelphia, in 1949. The studies helped determine that an ejection seat needed to have a screen pulled over the airman's face a split second before he blasted away the cockpit and ejected, usually by reaching up and pulling it down. As aircraft became faster and more powerful, the limits of human endurance became a major concern of the new field of aviation medicine. Research into supersonic and space flight in the 1950s focussed strongly on gravity forces (a 4G turn is body weight by four) and led to pilots wearing 'G-suits' to keep blood flowing to their brains so as to avoid black-outs – or even death.

Safe ejection

The pilot of a US Navy Crusader fighter 'bangs out' of his plane on the carrier *Franklin D. Roosevelt* in October, 1961, after his right wheel collapsed on landing and the arrester hook broke with the extra strain. Ejector seats have saved thousands of lives in military aircraft since the Germans introduced them towards the end of World War II. Britain's Martin Baker Aircraft Company is the world's biggest supplier, with some 70,000 supplied to 90 air forces, and some 6,700 lives saved since its first design in 1944. Explosive bolts simultaneously blow away the canopy and project the pilot and seat straight up with spine-compressing force. The newest Martin Baker NACES seat for the US Navy starts to deploy the parachute within half a second and includes an emergency radio and survival supplies. Every MB ejection survivor gets a tie and membership of an exclusive club.

AB
12
5357

Spy trial

Francis Gary Powers, the pilot of an American U-2 reconnaissance plane was shot down by a new surface-to-air missile at 65,000 feet over Sverdlovsk, Soviet Union, in May, 1960 (right). Powers' capture was a major military and political coup for the Soviets and encouraged a more aggressive stand towards the US. He was put on trial before a huge audience in the Hall of Columns in Moscow (far right, standing in dock) on charges of spying. Almost two years later he testified before Congress after being exchanged for the Soviet spymaster Colonel Rudolph Abel (above).

Cuba missile crisis

Perhaps the closest the world came to nuclear war was in October, 1962, when the US demanded the Soviets withdraw nuclear SS-4 ballistic missiles and Ilyushin Il-28 bombers from Cuba. The long-range weapons were far more than Cuba needed for its defence and directly threatened most major US cities. At first the Russians denied their presence, but high-flying U-2 aircraft flown by the CIA produced aerial photographs of nine sites being prepared. The Kremlin backed down and the weapons were withdrawn. Reconnaissance aircraft photographed the Soviet Kasimov freighter with 15 crated Il-28 fuselages on deck leaving Cuba in November, 1962. While the shooting-down of Gary Powers' U-2 in 1960 had dented the reputation of US aviation intelligence-gathering, the Cuban missile crisis restored it.

1963-1973

7 Speed versus size

The Sixties saw the culmination of the battle between military-inspired advances in speed and power and the civilian requirements for increased size and economy as airports expanded and the era of cheap, mass air travel dawned. In other words this meant supersonic Concorde versus the Boeing 747 Jumbo; the former seen taking off with its unique 'droop snoot' nose lowered on its maiden flight at Filton, Bristol, in April, 1969 (preceding page). Both aircraft had their roles, but Concorde also carried British and French hopes of remaining leading aviation nations. Britain's military aviation policy had been badly mauled in previous years as the government tried to decide how to replace its increasingly vulnerable V-bomber force. The successor supersonic TSR-2, publicly revealed in October, 1963, was perhaps ahead of its time and was cancelled in 1965. The government then ordered the end of the navy's big aircraft carriers, and vacillated over what navy plane to buy. The unique Harrier jump jet stepped in to partially save the day. The Vietnam War was proving to be the real-world military laboratory, however, with America's technological superiority demonstrated by aircraft such as this Sabre firing rockets at Viet Cong positions (right). The B-52 proved its utility as a conventional bomber, but it was the lowly Huey helicopter that emerged as the lasting image of the war (the late 1990s musical *Miss Saigon* used an initial 'S' like a helicopter).

RESCUE

Supply drop
In 1964, just before they poured their own troops into Vietnam, the US Army used helicopters to deliver supplies to South Vietnamese soldiers in a jungle clearing.

Vietnam call-up
Huge numbers of helicopters were required for the US Army in South Vietnam, including these Bell Chinook heavy-lift and light Sioux choppers in 1967.

Air Cavalry
Bell UH-1 Huey helicopters of the US 1st Air Cavalry Division became synonymous with the Vietnam War. During Operation Pershing in 1967 two soldiers guide in a second wave of troops for a 'search and destroy' mission on the Bong Song Plain. The French, using older and slower methods, had failed to regain their hold on Vietnam in the 1950s. The Americans hoped helicopter-borne troops would be quicker and more decisive.

Combat mainstays
The McDonnell Douglas F-4 Phantom II (above) was not only a lynchpin of the US Navy and Air Force in the 1960s, but also sold a total of 5,057 around the world. More than 1,000 are still flying. A prototype Dassault Mirage IV (left) became the nuclear-capable heart of France's new strategic air force, the FAS, in 1964.

The Jaguar
France and Britain formed a new joint company, SEPECAT, to build this fighter for both air forces (right).

Coke commercial

The Cold War was largely a matter of intense, intelligence-gathering as both sides sent ships and planes into international air and sea space to find out each other's military capabilities – with some rare levity. In NATO exercise 'Northern Wedding' in September, 1970, a British navy Phantom fighter from the carrier *Ark Royal* photographed crewmen on a snooping Soviet Tu-16 Badger bomber grinning and displaying a bottle of Coca-Cola (right). A formation of Soviet Sukhoi fighters (above).

British casualty

Probably the world's most advanced and complex aircraft of its time, Britain's TSR-2 (Tactical, Strike and Reconnaissance) bomber (above) was designed to replace the tactical nuclear V-bombers and the Canberras used for reconnaissance. Supersonic, built of titanium, the TSR-2 was the first plane to have terrain-following and side-looking radar. But initial engine problems and exaggerated cost claims prompted the new Labour government to cancel it in April, 1965. The 50 American F-111s the government decided to buy instead were also later cancelled.

Blackbird menace

After the Soviet shooting-down of Gary Powers' U-2 spyplane in 1960, the US determined never to be humiliated like that again. In secret at the Lockheed 'Skunk Works' in California it built the SR-71 Blackbird, intended to fly at Mach 3.3 (2,000mph) and at up to 120,000 feet so that no missiles or planes would ever catch it. The aircraft went into service with Strategic Air Command in January, 1966. It provided valuable intelligence to Israel in the 1973 Yom Kippur War.

N AM
JET CLIPPER AMERICA

Wide-body

Passengers in the Boeing 747 were seated in rows of 10, with two aisles, in-flight movies, a full range of meals, an upper deck, and even a spiral staircase. With a gross weight including fuel of 400 tons, and able to carry 400 passengers more than 5,000 miles, the 747 became a little world of its own, its passengers barely aware of the sensation of flight. The aircraft greatly worried airport managers, however, as four fully-laden 747s arriving within 15 minutes could mean 1,600 people, plus another 600 or so to greet them. Not to mention their luggage.

Crowd puller

A policeman stops the traffic as Britain's Concorde 002 (France's Concorde 001 flew at the same time) lands at Fairford, Gloucestershire, watched closely by a trailing Canberra chase-plane, after its brief maiden flight from Filton, Bristol, on 9 April, 1969 (right). Concorde has been a crowd-puller ever since. Designed for speed, the cigar-shaped fuselage of the 100-seat Anglo-French aircraft is long and slender, with a single aisle separating rows of just four seats (above). Concorde was not designed for the mass market, however, and the very expensive, first-class only ticket provided the finest wines and meals service .

160 VFC

Concordski
The Soviet Union's supersonic Tu-144 bore a close resemblance to the Concorde when it first appeared in the West at the Paris Air Show in 1971, but it struggled technologically (above). By the time it next appeared at the show in 1973 it had been reconfigured with retractable canards, or foreplanes, at the front of the aircraft to help control it at low speed. However, it crashed in a Paris suburb, an engine landing next to a house (left). The Tu-144 was never put into regular passenger service and after another crash was withdrawn in 1978.

Cold War giants

The first Antonov AN-22 transport aircraft to make its appearance in the West amazed visitors to the 1965 Paris Air Show by its unprecedented size (above). With a wingspan of 211 feet and length of 190 feet it was designed to carry complete mobile power stations to Siberia, one or two 64-ton T-62 tanks or up to 724 passengers in a stretched version. From the US in 1968, however, came Lockheed's C-5A Galaxy, longer at 247 feet and with a wingspan of 222 feet (right). It became the strategic heavy lift workhorse of the US Air Force.

12

Definitely a Jumbo

Proving the point of calling big aircraft jumbos, a Transmeridian Air Cargo Canadair CL-44 aircraft flew 14 tame elephants from New Delhi to Stansted Airport, England, in January, 1972. The alternative to flying was a long sea voyage. Twelve of the elephants were destined for Billy Smart's Circus, two for Fossett's Circus, Northampton. Until the AN-22, the Canadair CL-44 was the largest transport plane in the world.

BEA

Hands on and hands off

In June 1965 Eric Poole (left) and Tommy Atkins (above), pilot and co-pilot of a British European Airways Trident airliner, could show how they made the first hands-free automatic-pilot landing on a passenger-carrying flight. In March, 1970, however, a strike by British airport employees forced passengers to carry their own bags from their BEA plane (left).

Up and down

The growth of airline passenger business in the 1960s meant vast amounts of baggage to be dealt with at airports. In the US automatic facilities were built to move baggage to carousels.

Round and round

Five DC-8 airliners form a symmetrical pattern around a new United Air Lines concourse connected to San Francisco International Airport in the early 1970s. Passengers boarded and disembarked through 'Jetway' telescopic corridors extended and retracted as needed between plane and concourse.

Moving walkway
A huge experimental moving walkway in the Trans World Airlines terminal at John F. Kennedy Airport in New York presents striking lines and hints at why many find airports such alienating places. Some airports developed mobile lounges, others automatic trains.

Chopper rescues
Helicopters became ever-more useful during the 1960s and 70s, particularly when it came to saving lives. In Switzerland, a helicopter arrived twenty minutes after a car plunged into the fast-running River Rhone (far left); a crewman was lowered to the car, and the driver and his passenger were lifted to safety (left).

Six-Day War

Three Egyptian Tu-16 bombers lie smashed on the ground at their airbase (above), at the start of the 1967 Six-Day War, which was a triumph for the Israeli Air Force. In the first day alone it put 25 airfields in Egypt, Syria and Jordan out of action and destroyed 350 aircraft, all but 24 on the ground, for the loss of 19 Israeli aircraft.

Close cooperation

A Sud-Ouest Vautour bomber of the Israeli Air Force flies low over Israeli armour moving into position on the southern front during the 1967 Six-Day War. Once Israeli Mirage pilots had quickly established air superiority, IAF bombers smashed retreating Egyptian forces at the Gidi and Mitla passes in the Sinai and then, on the fourth day, hit Syrian troops in the north until the Golan Heights were gained.

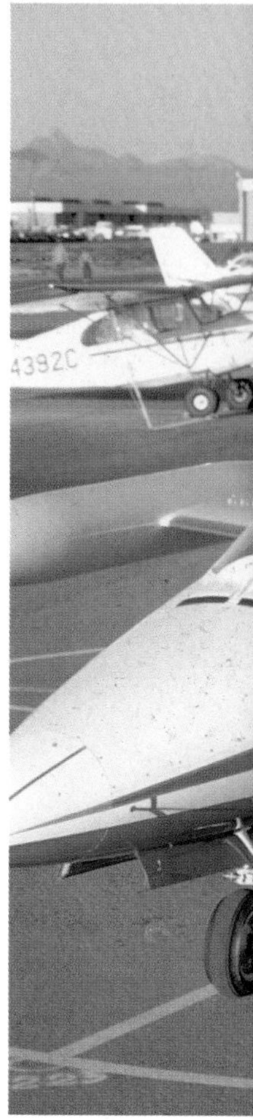

Plane garage

Keeping a Cessna plane in the garage gave Ralph Hannum a certain status when he had a house built for his family at Downington, Philadelphia, in 1965. Since then 'air parks' have sprouted throughout the US, providing aviation facilities for communities of people who fly their aircraft for business and personal use.

Personal transport
David Murdock and his wife and two children in front of a Lear Jet at Scottsdale Airport, Arizona, in 1973. Crammed with small aircraft, the airport grew rapidly in the 1970s as American businesses moved out of the 'Rust Belt' and towards the warm, sunny south west. Lear Jets and other executive planes allowed them to keep in regular contact with their customers and suppliers hundreds of miles away.

Centre to centre
One of aviation's cherished dreams, an aircraft that could fly transatlantic from city centre to city centre, took a step closer to reality in April, 1969. An RAF Harrier jump jet carried Squadron Leader Tom Lecky-Thompson most of the way from the top of London's Post Office Tower – via a spectacular take-off from a coal yard at St. Pancras Station and air-to-air refuellings – to the top of the Empire State Building, New York, in six hours 11 minutes. He was beaten by an hour by a Royal Navy pilot in a faster Phantom, but the stunt demonstrated the promise of Vertical Take off and Landing (VTOL).

Terrorists strike

One of three airliners blown up by Palestinian hijackers at Dawson's Field in Jordan in September, 1970. A fourth was blown up in Cairo. An attempt on an El Al plane failed, and hijacker Leila Khaled was held in London. She and three other terrorists held by the Germans and Swiss were swapped for 56 passengers.

egistrement
D27
Heathrow
Heathrow
Heathrow
Heathrow

1973-1983

8 Fly-by-wire and the mass market

The separation of trends in mainstream military and civilian aviation accelerated during the 70s and early 80s. This was the decade that introduced the combat jets that were at the heart of the world's defences at the end of the century: the British-German-Italian Tornado, Soviet MiG-29, French Mirage 2000, and American F-14, F-15, F-16 and F/A-18. This last (right) lost out to the F-16 in the contest for the US Air Force's new lightweight fighter, but the Navy took it on and introduced the 'glass cockpit' of electronic screens rather than dials. The aircraft were the first designed not to fly in the traditional sense, with most of their mechanical flight control systems between pilot and elevons and rudder replaced by electronic wires connected to computers that made them automatic and much more responsive than if under human control. 'Fly-by-wire' made flying easier and the aircraft lighter but, by enabling sharper 'G-force' turns, it also revealed the probable outer limits of manned flight. Civilian airliners were built more cautiously, and would not see fly-by-wire controls for more than a dozen years. For the airlines, their economics were much more important, with airline deregulation beginning in the US in 1977 and in Britain the likes of entrepreneur Freddie Laker introducing cut-price fares on his Skytrain. With mass travel exploding, there was insufficient aircraft capacity, union unrest, and jams at airport check-in desks, such as Heathrow's Terminal Three in 1978 (preceding page). Extra security checks slowed everything down. But the alternative? Mid-east terrorism infected aviation in the 70s and 80s, and while there were still hot wars such as the 1973 Arab-Israeli war, the last throes of Vietnam, and the Falklands War, civil aviation was itself becoming a battleground.

307
VFA-82
07

Night departure

A Pan American 747 at New York's John F. Kennedy airport prepares to push back for a night flight. PanAm saw a 59 per cent increase in New York-London passengers when it put 747s into operation the previous year. World airline passenger travel more than doubled over the decade.

Airbus challenge
With the backing of the French, German and, initially, the British governments, a consortium of European companies challenged American dominance of the wide-bodied market in 1972 with the first multinational airliner, the 290-seat Airbus A300B. The world's first twin-engine, wide-body jet transport, the aircraft went into service with Air France in 1974 and by 1979 had broken into the American market with airlines such as Eastern (right). By 1980 Airbus had 26 per cent of the world-wide market by value.

EASTERN

Yom Kippur War

An Israeli A-4 Skyhawk levels for an attack on Syrian positions on the Golan Heights, watched by an Israeli soldier, on the third day of the 1973 Arab-Israeli war. Unlike the 1967 war, Israel did not have command of the air in the opening days, particularly against Egypt. The war was keenly analysed by the US and Soviet militaries because it pitted the initially-successful Soviet-supplied SAM-6, SAM-7 anti-aircraft missiles and ZSU-23 four-barrelled radar-guided guns against new US-supplied Skyhawks, F-14 Phantoms, Sidewinder air-to-air missiles and Hawk surface-to-air missiles.

The Entebbe raid

In one of the decade's most celebrated military actions, in July, 1976, an elite force of Israeli commandos flew a 5,000 miles round-trip to rescue more than 100 mostly Jewish passengers from a hijacked Air France plane held hostage in Entebbe, Uganda. The commandos arrived in four C-130 Hercules transport planes (one of which is seen right) in the dead of night immediately behind a British cargo plane, confused Ugandan security guards further by driving out of one plane a Mercedes and two escorting Land Rovers identical to those of dictator Idi Amin, then shot up the control tower, blew up 13 Ugandan MiG fighters and killed 35 Ugandan soldiers and all 13 terrorists. One hostage left in a Ugandan hospital and one commando – the raid leader Yoni Netanyahu, brother of Israeli politician Benjamin – were also killed, but all the rest were saved and when they arrived back at Tel Aviv they were heroes (above).

4X-FBQ
420

All Aboard!

Sir Freddie Laker at the launch of his 'Skytrain' enterprise in September, 1977 (right). Skytrain was a hugely popular no-frills, low fare operation which worked on a 'first-come, first-served' basis. In 1984, entrepreneur Richard Branson, joined the cut-price airlines war by launching Virgin Atlantic, offering a one-way fare from London to Newark, NJ, for £99 on its single leased Boeing 747 (above). Branson, whose mother was an airline stewardess in the 1930s, went on to be one of the most adventurous names in aviation.

Speedy first class, slow steerage

While the mass of airline travellers fumed and fretted at the regular economy desks, Concorde passengers were whisked through a special check-in desk at Heathrow airport (above). They did, of course, pay up to seven times more for their first class supersonic seats. Getting Laker Skytrain tickets at Gatwick airport, near London, in 1978, could be a very slow process. Some people decided that queuing horizontally, preferably while unconscious, seemed to while away a great deal of time (right).

The Fun airline

Southwest Airlines of Texas was a pioneer of the concept of 'sex sells seats'. Stewardesses in 1972 wore hot pants and kinky leather boots (right) and were selected as much on their legs and face as their skill at serving drinks such as 'Passion Punch' and 'Love Potion' to their passengers. In 1982 the airline dressed up its short-haul planes as killer whales (above).

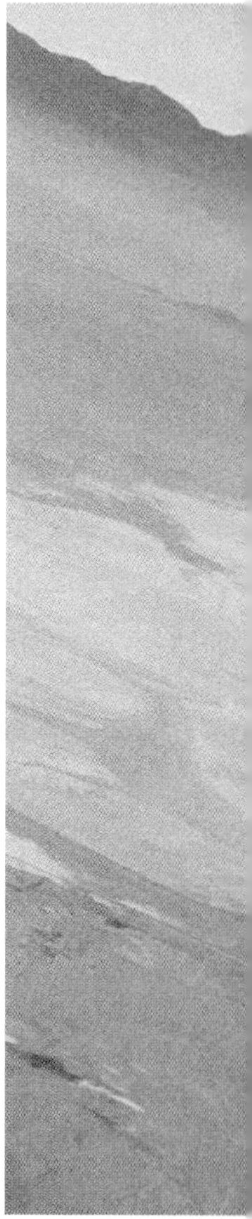

Blast off

Astronauts John Young and Robert Crippen blast off in the first space mission of STS-1, the Orbiter Columbia Shuttle, in April, 1981, the first attempt at a re-usable space craft (above). Both were experienced military pilots.

Space-glider

Hardly a rocket ship, but hardly an aeroplane either, the Space Shuttle Enterprise flies free from its special 747 'mother-ship' in 1977 to test its ability to glide down to Rogers Dry Lake in California, once it breaks back into the atmosphere from space (above).

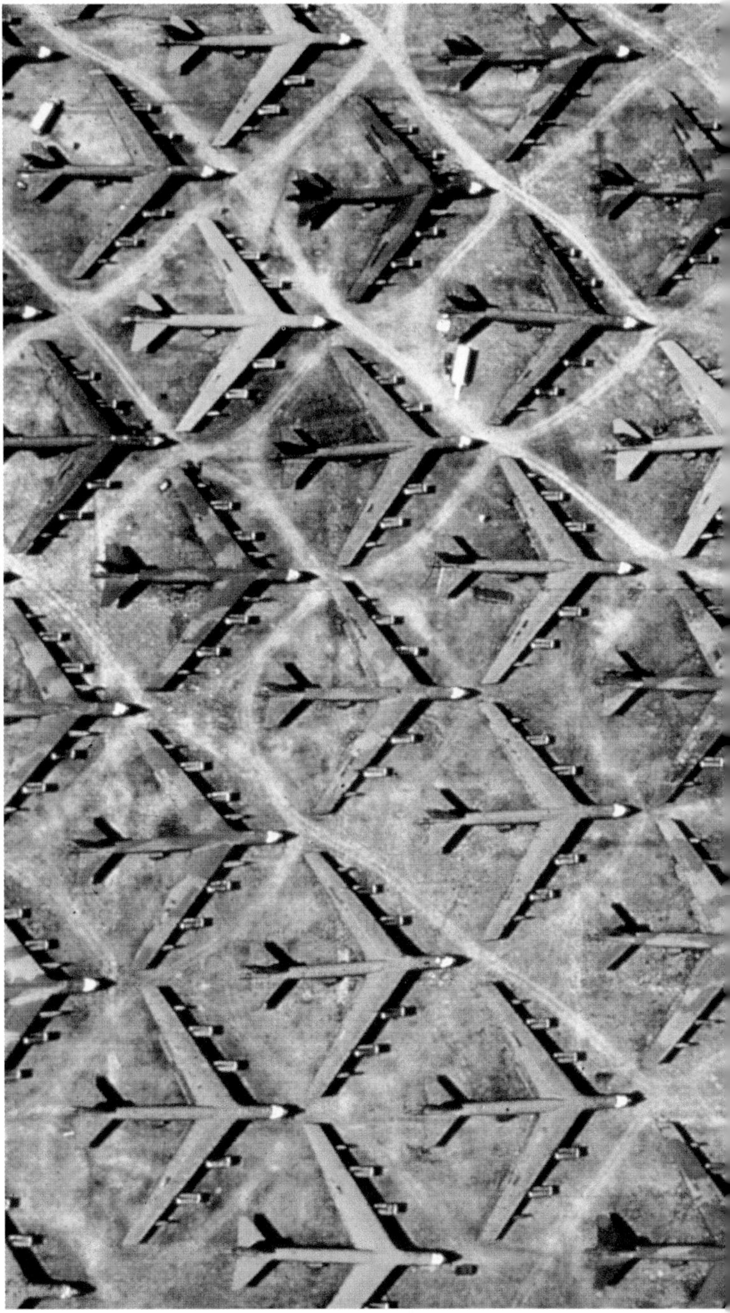

Aircraft graveyard
An aerial view of an aircraft storage and scrap yard at Davis-Monthan air base in Arizona produced an interesting pattern in the late 1960s. Most of the aircraft are still-useable B-52 bombers, parked in the dry desert where rust cannot eat them away. The 'graveyard' is used regularly by the Pentagon and airlines to store planes until they are sold or brought back to service.

Airborne control
Commanding battles, identifying enemy aircraft and controlling friendly ones at ever-greater distances became a vital part of air power in the 1970s. In 1973 the USAF received the first of several Boeing 747-200Bs radically modified to be E4-A command posts (above), equipped with consoles and accommodation for senior officers to direct operations in the event of nuclear attack.

Early warning
The first AWACS aircraft, a converted Boeing 707 used to identify hostile aircraft hundreds of miles away with its huge rotating circular radar scanner, joined the USAF in 1977 (above). The British could have done with the US Navy's carrier-borne E2C Hawkeye early warning (AEW) plane (left) during the Falklands conflict.

Preparing for battle
British Royal Marines check their weapons in the hangar of the aircraft carrier *Hermes* on their way to battle with Argentinian forces occupying the Falkland Islands in April, 1982. On the deck above crewmen prepare for night operations by the RAF and Navy Harriers. The Falklands, 8,000 miles away in the South Atlantic, heavily tested British air power, vital to protect and secure sea and ground operations to recover the islands.

Exocet threat

Argentinian Super-Etendards, similar to these French navy aircraft, posed the deadliest threat to the British fleet in the Falklands conflict because several of them carried French-made Exocet sea-skimming cruise missiles. The Royal Navy had little time to react to the Super-Etendards flying at sea level at 650mph and the Exocets launched 20 miles away at 700mph. The destroyer HMS *Sheffield* was the first ship sunk by an Exocet. Another, the *Atlantic Conveyor* container ship, was mistaken for a British aircraft carrier, but its loss was bad enough – five of six vital Chinook helicopters were on board. Rapid-firing guns and clouds of disorientating metallic 'chaff' were a ship's only close-in direct defence. Fortunately, Argentina only had a few Exocets and the French ensured they got no more.

Harrier protection

British Harrier jump jets – 28 navy Sea Harriers and 14 RAF GR3 strike Harriers – were all that protected the British task force from attacks by some 100 Argentinian Skyhawks, Mirages, Daggers, Pucaras and Super-Etendards. The British had no long-range early-warning radar, but the Argentinian pilots were at the limits of their range and bravely pressed their attacks knowing they had no time to engage in dogfighting. The Harriers shot down 20 aircraft, with another three probables, without loss.

of the

1983-1993

9 Military and civil worlds collide

Cracks started to appear in the Iron Curtain at last, partly because the Soviet Union just could not afford to keep up with huge technological advances in American space and aviation equipment. One such crack was caused by the flight of German teenager Mathias Rust in a single-engine Cessna from Finland to Red Square, Moscow (right) through supposedly one of the world's most sophisticated air defences. It was the decade that saw 'invisible' Stealth planes and satellite-guided cruise missiles used in war for the first time, as almost the whole panoply of Nato's military might was brought to bear on Iraq for invading Kuwait in 1990. But it was also the decade that brought the military and the still-growing world of civil aviation into tragic collision. In 1983 two Soviet fighters shot down Korean Air Lines Flight 007 over Sakhalin in the Soviet Pacific, apparently believing it to be an American reconnaissance plane. In 1988 the Americans did much the same thing when the guided missile cruiser *Vincennes* shot down an Iranian Airbus over the Gulf. But when a bomb exploded on Pan Am Flight 103 over Lockerbie, Scotland in 1988 (preceding page), killing 270 people, there was nothing accidental about it. Europe's Airbus began to make serious inroads into Boeing's domination of the world of large passenger planes. But the 80s also saw lighter materials allow major new developments in personal flying machines – hang-gliders, microlights, paragliders, and even pedal-powered planes.

TWA hostage

Unshaven and exhausted, TWA pilot John Testrake looks out of the flight-deck window of his hijacked plane in Beirut as a Lebanese Shi'ite Moslem gunman brandishes his pistol (left). The hijacking of the Boeing 727 airliner started in Rome in June, 1985. The gunmen, demanding the withdrawal of Israeli troops from Lebanon and release of some 700 Shi'ite prisoners held in Israeli jails, directed the 153 passengers and crew to fly to Algiers, Beirut, Algiers and then Beirut again. The gunmen, one of whom carried a machine gun when he emerged (above), released more than 100 passengers, but killed a US Navy diver and dumped his body on the Beirut tarmac. President Reagan ordered an aircraft carrier to the area, but before the US could retaliate Israel announced it would meet the terrorists' demands. The gunmen escaped.

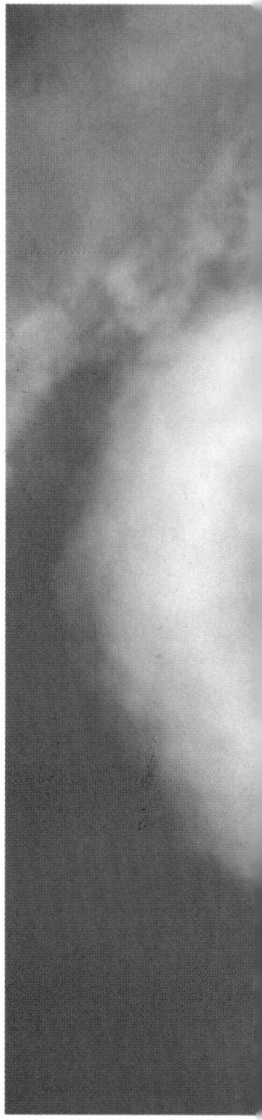

Eurofighter points ahead

Europe's largest and most expensive collaborative military aircraft, Eurofighter, first took to the air as British Aerospace's Experimental Aircraft Programme (EAP) demonstrator in August, 1986 (above). Over the next few years it developed as EFA, the European Fighter Aircraft, to include Germany, Italy and Spain, while France dropped out to develop its own Rafale fighter. A total of 620 have been ordered, with Britain getting the largest number.

Rafale spins

A French Rafale fighter, air-show smoke streaming from its wings, demonstrates how tight it can twist and turn (above). The Dassault Rafale (Hurricane) is intended to replace almost every fighter in the French air force and navy and was designed to be lighter and less powerful than Eurofighter partly to accommodate it on aircraft carriers. The first Rafale A prototype flew in July, 1986.

Air show tragedy

Air shows attract millions of people around the world every year, and in some countries such as Britain are ranked second only to football as a spectator sport. While safety regulations have increasingly separated spectators from flying aircraft, there have been some disasters, the worst in recent years being at Ramstein Air Base in Germany in 1988. There the Italian Air Force's 'Frecce Tricolori' aerobatic team collided (right) and crashed among the crowd, killing 70 people and injuring more than 450, some of them seriously.

Lockerbie wreckage

A rescue worker carries a rope out of a huge hole gouged in the village of Lockerbie, Scotland, by the impact of Flight 103, blown up by a bomb as it flew from Frankfurt to New York via London in December, 1988. Twelve people were killed on the ground together with the 258 passengers and crew. More than 40 homes were destroyed.

Aviation sleuths
It took many months of painstaking work, but the downed Boeing 747 of PanAm Flight 103 was partially reconstructed in order to see exactly how the bomb exploded and brought down the plane. Twelve years later, after a nine-month trial, only one of two Libyans arrested for the crime was found guilty of murder by a Scottish court.

LIFE VEST

Computerised Airbus

The flight deck of the new Airbus A320 in 1987 (left) reveals the world's first commercial aircraft to make full use of 'fly-by-wire' electronics, already common in new military aircraft for a decade. Instead of the centred 'joystick' and dials of older planes such as the RAF Vulcan (above), the A320's have 'sidesticks' for pilots to control the aircraft, and colour screens provide flight and navigational information. Seven on-board computers replace mechanical cables to guide all flight-control functions. The computers mean the aircraft cockpit is only designed for a flight crew of two.

Low level terror

While American bombers struck from mid-level height (12,000-feet plus), Britain's Tornado GR1's attacked Iraqi targets as they had long trained for in the Cold War – at low level, 100 feet or less (above). While that got them under Iraqi radar, less expected was the enormous barrage of anti-aircraft fire thrown haphazardly into the air, hinted at in this night scene of Baghdad as a bomb explodes (right). Tornado pilots described it as a terrifying 'wall' through which they hardly expected to survive. A myth grew that most, if not all, Tornado losses in the Gulf War were due to low-level tactics, and after a few days the Tornados switched to medium-level bombing. But this was found not to be so, and the RAF returned to practising low-level techniques after the war.

Gulf War

US Army Apache helicopters (right) launched the Allied offensive against Iraq in the Gulf War in the early hours of 16 January, 1991, when they crossed over the Saudi-Iraq border to destroy two Iraqi early warning radars that sat across the main route for US electronic warfare planes. The Apache's Hellfire missiles, 70mm rockets and 30mm cannon each have a striking power worth a dozen tanks and have given several armies new-found air power, including the British Army in late 2002. An air attack on Iraqis fleeing Kuwait at Mutla Ridge (above), however, was so devastating (1,400 vehicles destroyed, some 250 people killed) that it had a direct impact on the end of the war. Pictures of the 'Road of Death' so horrified public and politicians at home that Allied attacks were ended, arguably allowing Saddam Hussein's elite Republican Guard to escape and him to continue in power.

Stealth

Two US F-117A 'Nighthawk' Stealth fighters make a strange and ominous sight as they wait for the cover of darkness to do their work in the Gulf War (above). The F-117s were built at Lockheed's secretive 'Skunk Works' in California and first flew in 1981. The fly-by-wire planes were the first to use flat sides and sharp angles to deflect enemy radar beams and reduce the size of their 'signatures' on radar scopes, to the point where they were virtually invisible. They were the first aircraft to bomb Iraq's heavily-defended command and control bunkers, using precision-guided Paveway II bombs (left).

U.S. AIR FORCE

Cruising

An air-launched cruise missile (ALCM) drops from the bomb bay of a B-52 bomber before flying to its target up to 1,500 miles away. The B-52/ALCM package complemented the US Navy's Tomahawk cruise missiles, seen flying at tree-top height down streets in Baghdad on the way to their targets, and gave the venerable bomber a new lease of life. Since the Allies had total air superiority, B-52s could also drop thousands of 'dumb bombs', directly over targets.

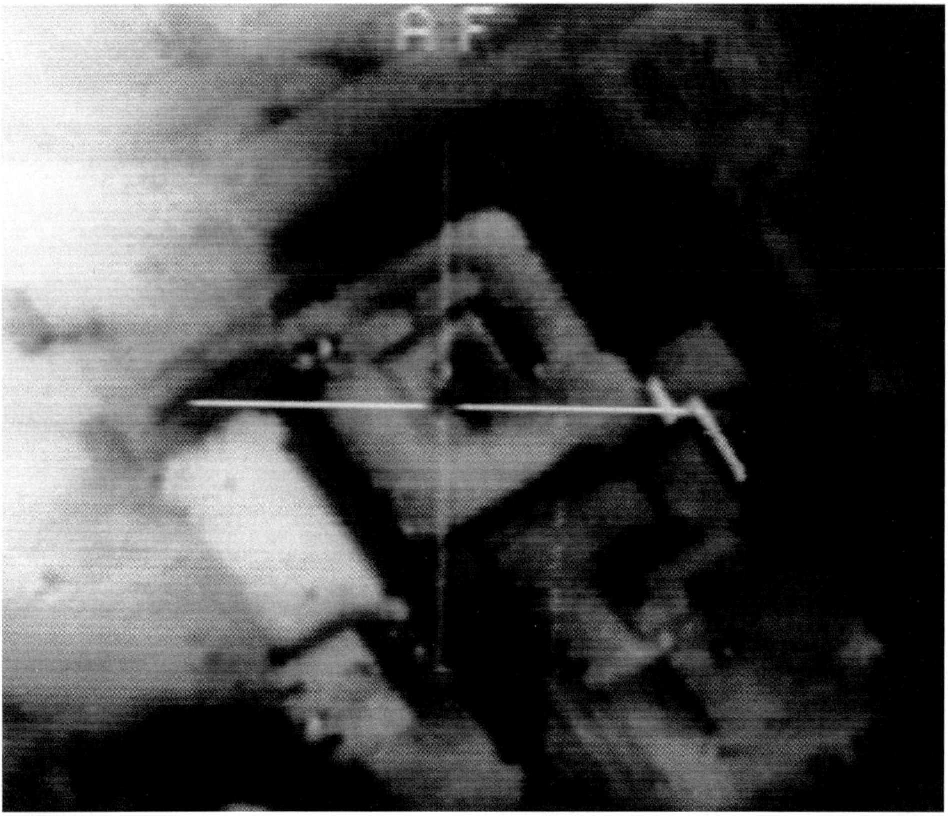

Total focus

Days after his return from the Gulf War in 1991, Capt. Mark Miller of the USAF's 338th Tactical Fighter Wing focuses forward through the Head Up Display of his F-16 at Hill Air Force Base, Utah (right). The intense concentration required of a combat aircrew is rarely appreciated by those who have never experienced it. One of their jobs in battle is to 'lase', or use a toggle in the cockpit controlling a guidance laser, in order to keep the cross-hairs onto the target, so that the laser-guided bomb can track down onto it. The image the Weapon Systems Operator (navigator) sees is shown in this recorded view through a video camera mounted in the nose cone of a Paveway bomb aimed at the roof of the Iraqi military's main communications building in Baghdad on 19 January, 1991, the opening night of the Gulf War.

Surrender
Iraqi troops surrender to helicopter-borne French special force troops on 26 February, 1991, half-way through the 100-hour ground war. The speed of the Allied advance, after an intense air campaign lasting more than a month which left the Iraqis battered, confused, starving and thirsty, allowed Allied soldiers to make many such easy captures. Some groups of Iraqis surrendered directly to hovering helicopter gunships.

1993-2003

10 Aviation reigns supreme

The decade leading to the centennial of the Wright Brothers' first manned flight was a sombre one for the most part. While there was much to celebrate – as Argentine Air Force personnel did to mark the 30th anniversary of the first aircraft to land at Marambio base, Antarctica, in March, 2000 (preceding page) - the events of 11 September, 2001 in America underlined the awesome power of aviation. The global impact was unprecedented: the effect on world trade, the subsequent aerial assault on al Qaeda and Taliban bases in Afghanistan, and the political and religious tensions stirred up. Aviation is the tool by which power is increasingly exercised around the world. Over the decade armies developed air-mobile rapid reaction forces and navies installed longer-range missiles. Precision-guided bombing went from 10 per cent in the Gulf War to 80 per cent in Afghanistan. The use of unmanned aerial vehicles (UAVs) rose rapidly as the decade closed, posing questions about the future of manned flight itself. But while military technology soared ahead, defence budget cutbacks required more 'off the shelf' civilian equipment purchases by the military and that economic civilian methods of production be used. The aerospace industry restructured itself, with yet more company mergers and businesses becoming truly global in size and scope. Humanitarian relief flights and helicopters saved ever more lives, such as this family trapped in flood waters in Mozambique in March, 2000 (right).

Emergency dash

An air ambulance, with people clinging to the skids, takes off from a street in Karachi, Pakistan, carrying the body of 42-year-old Murtaza Bhutto, estranged brother of the Pakistan Prime Minister, who was killed with six of his followers in a gun battle with police in September, 1996. Air ambulances have become more widespread around the world as traffic conditions worsen and the public have higher expectations of medical care.

Drugs bust

An anti-drugs police officer directs a US-supplied Black Hawk helicopter in Colombia after using it to swoop on a cocaine laboratory in the north-east Catatumbo area in May, 2000. Air surveillance and transportation have an increasingly important role in fighting the drugs cartels in Colombia.

Black Hawk Down

The wreckage of an MH-60 Black Hawk helicopter in Mogadishu, Somalia marks a low point for US special forces in 1993 . The 160th Special Operations Aviation Regiment (Airborne) attempted to take US Delta Force and Ranger troops into the congested city at rooftop height in daylight to capture Somali warlord leaders. Two helicopters were downed by rocket-propelled grenades and the operation turned into a fighting retreat. Eighteen US soldiers and 500 or so Somalis died in the 15-hour battle.

Nightstalkers don't quit

This was the motto of the 160th, a night-flying unit. Four Black hawks look menacing in silhouette (above).

Tomahawk strike
A US Navy warship launches a Tomahawk cruise missile against a target in Iraq during Operation Desert Fox in 1998. Tomahawks, which can also be fired from submarines, have given the US and British navies a major new role in spearhead air operations, striking high-value targets such as command bunkers and communications centres, and helping to pave the way for manned strike aircraft.

Stealth bomber

Whether in silhouette (right) or from the side (above), the tail-less, bat-wing Northrop B-2 stealth bomber makes an amazing and almost alien sight. First flown in July, 1989 and costing $500 million each, the B-2 is the most expensive aircraft ever. It is built with composite materials to absorb radar waves, rather than deflect them like the F-117A, and its engine exhausts are located on the top rear to minimise heat detection from below. Its operational debut came in March, 1999 when two B-2s flew a 31-hour, 16,000-mile non-stop mission from Missouri to drop thirty-two 2,000-pound satellite-guided bombs on Serbia. B-2s flew even longer missions from the US to strike Afghanistan in 2001.

Fast roping

British Royal Marines descend rapidly down ropes from a Sea King helicopter onto the deck of HMS *Ocean* during exercises off the coast of Sierra Leone in May, 2000. *Ocean* is Britain's first purpose-built helicopter carrier, and was a major base for marines in the Indian Ocean preparing to go into Afghanistan in 2002.

Aero-copter

US Marines jump from a unique aircraft that they have been trying to put into service for years – the MV-22 Osprey, a twin-engine aircraft that rises vertically and then tilts its stubby wings to fly forward like a plane. It is a key part of Marine Corps doctrine to lie off, unseen, much further away from a coast than ever before, and then fly in fast to seize on-shore targets. It has had technical problems, however, and has long been a source of political contention between the US Congress and the Pentagon.

L
ROYAL NAVY
ROYAL NAVY
003

European carrier power

Two Royal Navy Sea Harrier F/A-2 jump jets hover as they prepare to land on the carrier *Illustrious* (left). The Harriers, usually heavily laden with fuel and weapons, use a small ski-ramp to help them take off, but land by hovering alongside the ship and then move sideways to drop down vertically onto the deck. Britain is planning to replace its three 19,000 tonne carriers with two new conventionally-powered carriers at least twice as large as *Illustrious* and to fly new Joint Strike Fighters from them. France is currently committed to the Rafale fighter and the nuclear-powered carrier *Charles de Gaulle* (above), and is looking to fund a second large carrier.

415
VFA-86

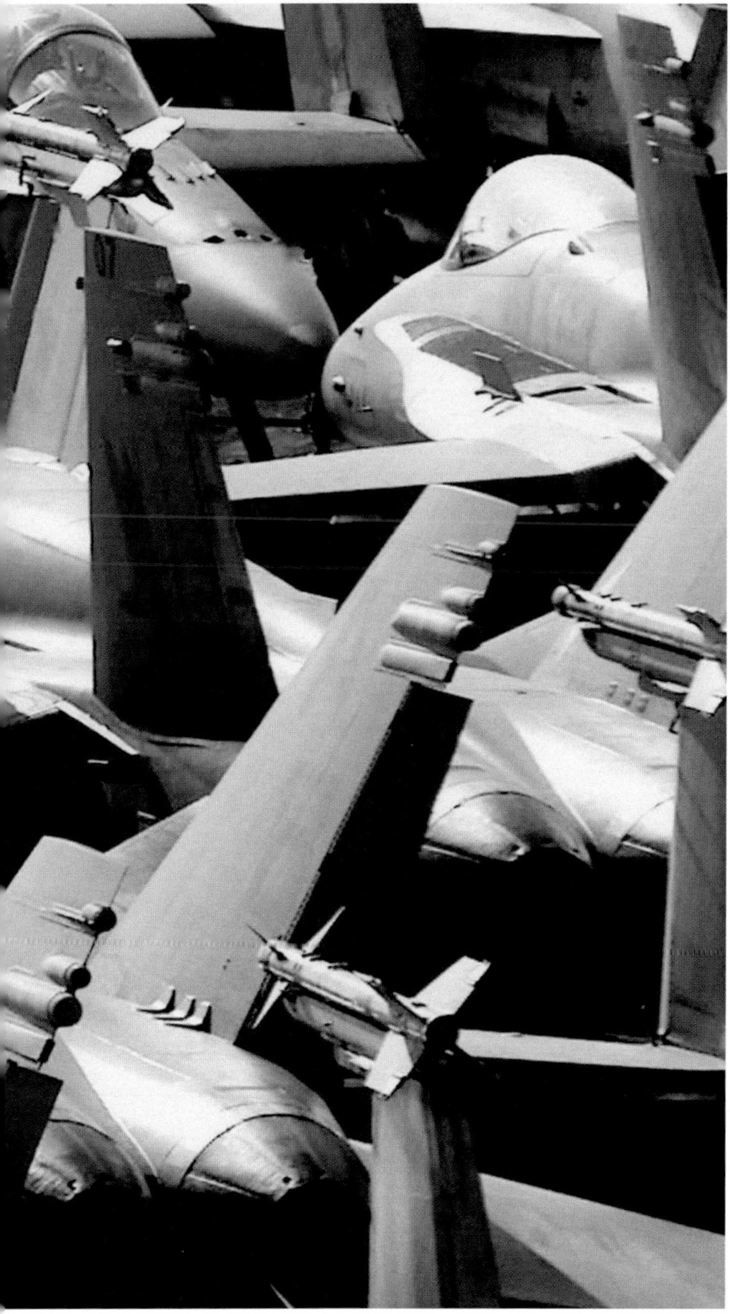

Sea of planes

As if with nowhere else to walk, a US sailor walks across the wing of an F/A-18 fighter on board the carrier USS *Theodore Roosevelt* in the Indian Ocean in December, 2001. Most of the 70 combat jets on board were involved in operations over Afghanistan.

The sound barrier

First experienced by Chuck Yeager in 1947, the sound barrier's visual impact was photographed by John Gay of the US Navy in July, 1999 as a US F/A-18 went through it close to the aircraft carrier *Constellation* in the Pacific. It is believed that the bubble effect was caused by the aircraft outrunning all the pressure and sound waves ahead of it as it exceeded 741 mph, the pressure change condensing water in the air so that it became a vapour bubble out of which the plane emerged..

Concorde aflame
A blazing Air France Concorde, taking off from Le Bourget airport, Paris in July, 2000, is caught in this video still frame taken by a passing motorist. Seconds later the plane crashed, killing 100 passengers, nine crew and four people on the ground. An investigation concluded that a piece of metal had dropped onto the runway from an earlier aircraft, shredded a Concorde tyre and thrown up debris into the Concorde's fuel tank, which then caught fire. After modifications Concordes were allowed to resume transatlantic service in late 2001.

汽車防銹中心
車中寶
Castrol
Coca-Cola
MOTUL

Last look
Cathay Pacific's 'Spirit of Hong Kong '97' Boeing 747 made a last landing between apartment blocks at Hong Kong's Kai Tak airport in July, 1998, just hours before the airport closed and the new $20 billion Chek Lap airport opened for commercial traffic on reclaimed land further away. Landing at Kai Tak long presented one of the world's closest physical relationships between a city and the aircraft that served it.

World's biggest plane
The Antonov 225 was built in 1988 as a booster to the Buran space shuttle, cancelled in 1991. Only one now exists, but it may yet be produced in quantity.

ANTONOV 225
410
300
413
U.S. AIR FORCE

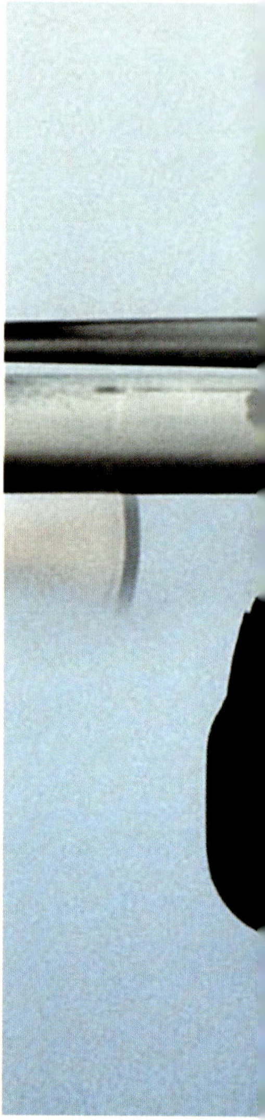

Afghan aid
An Uzbek soldier stands guard at Termez, Uzbekistan, as 40 tons of humanitarian aid from Denmark is transhipped for Afghanistan in November, 2001 (above). Landlocked and remote Afghanistan could not have been rapidly reached without aircraft, either to bring aid or to attack it.

11 September, 2001 (previous spread)
The date when hijackers seized four airliners in the US, flew one into the Pentagon in Washington, crashed another in Pennsylvania and flew two into the World Trade Center in New York. The impact of the second crash is shown, before both buildings collapsed. In all some 3000 people were killed on this day.

Flying tank
The USAF brought in its AC-130U Hercules to strike al Qaeda and Taliban positions in the mountains. This 'Spectre' gunship (above) carries an electronically-guided 105mm howitzer on its port side, along with a 40mm Bofors cannon and a 25 mm cannon. The 'U' model can engage two ground targets simultaneously and carry twice the ammunition of earlier models.

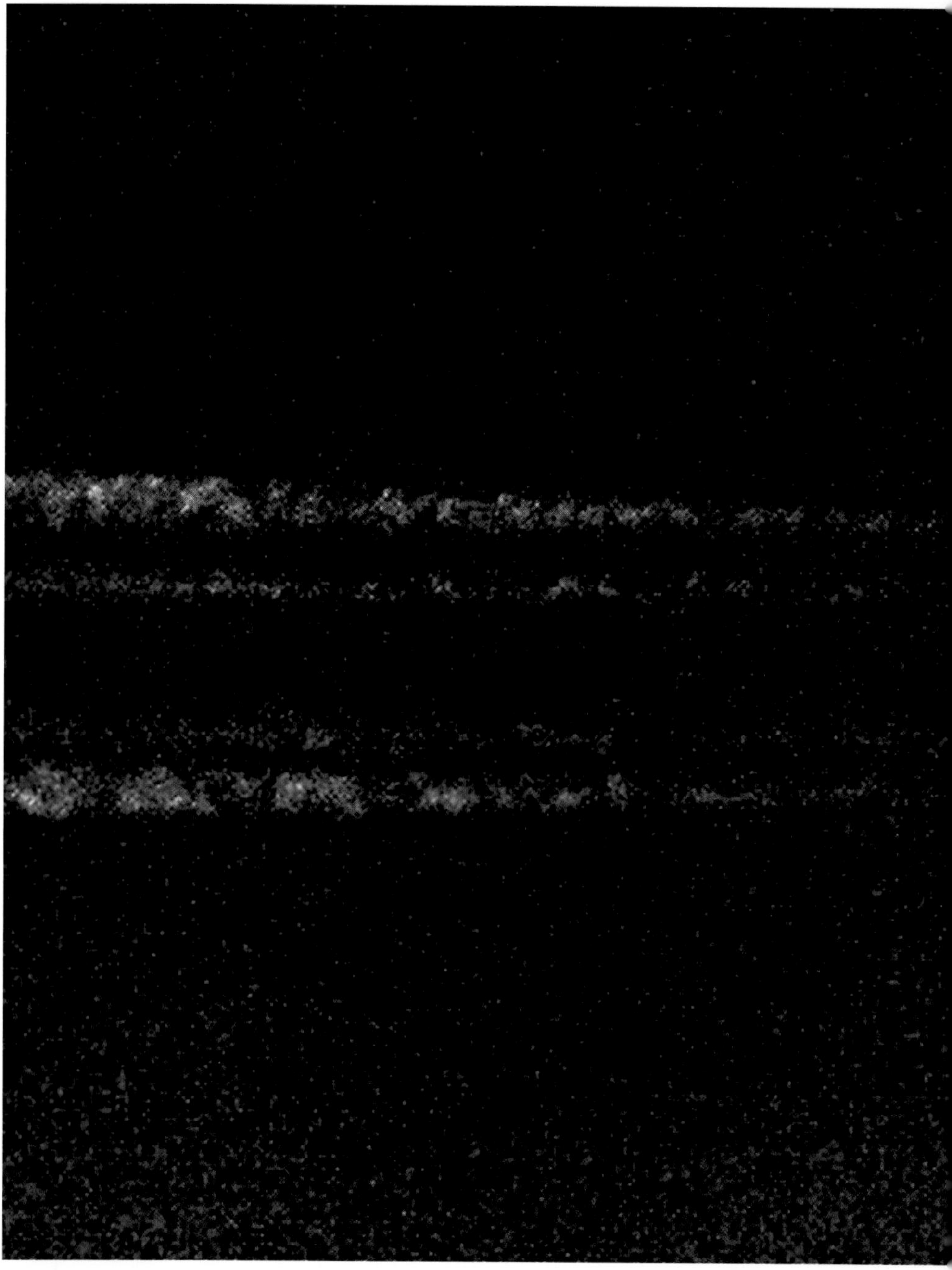

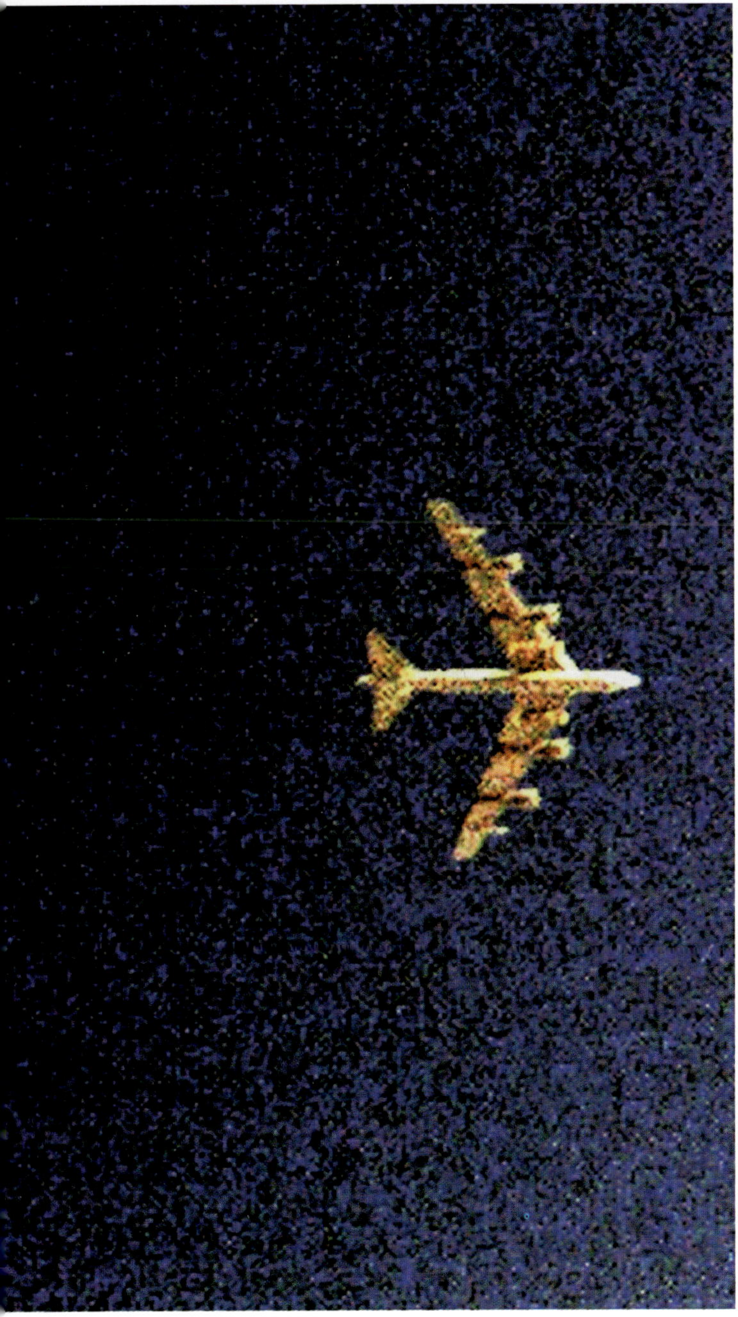

High flyer BUFF
Against a clear blue sky, a giant B-52 flies tens of thousands of feet above Afghanistan after releasing its bombs on front line Taliban positions near Bagram Airport in November, 2001. The elderly BUFFs (Big Ugly Fat Fellas in USAF parlance) have been in service for 50 years and are considered so useful the USAF plans to keep them for a further 30. Their parts are constantly renewed.

Pilotless planes

Hunter UAV

The British Phoenix UAV

Hunter UAV

Predator UAV

Global Hawk UAV

In February, 2002, in the first strike mission of a Predator UAV, flown by a CIA operative watching a TV monitor, was fired a Hellfire anti-tank missile at an al Qaeda target. In the Balkans, besides US Predator and Hunter UAV surveillance planes, France, Germany and Britain also used UAVs. The US Global Hawk UAV completed a historic 23-hour, 8.600-mile flight across the Pacific to Australia in April, 2001.

9/11 fallout

The 11 September, 2001 terrorist attacks on America caused a significant decline in air travel and more than two months later unused airliners sit on the tarmac at the Southern California Logistics Airport in Victorville (left). Around the world many older aircraft were taken out of service waiting for business to pick up. Cut-price airlines boomed, but in Switzerland the struggling state-owned airline Swissair went bankrupt, with part of its logo taken down at Kloten Airport, Zurich, in March, 2002 (above). The airline partially re-emerged minutes later, however, as Swiss, a merger with the regional airline Crossair and a partnership with American Airlines.

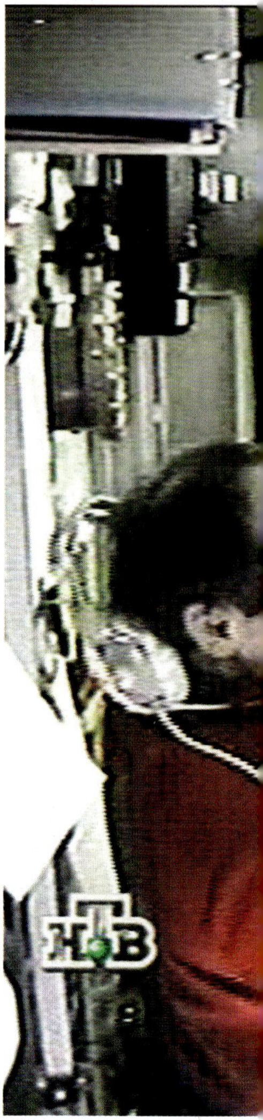

Mini space shuttle

Attuned to the money-making prospects of space tourism, in March, 2002 a Russian company unveiled a full-scale model of a three-seater spaceplane it hopes to be launched from the back of its record-breaking M-55 Geofizika altitude plane (at rear) in 2005. A £10 million prize – the X Prize – was offered in St Louis in 1996 for the first privately-funded team of engineers to design, build, and fly a reusable sub-orbital spacecraft, which must fly twice within two weeks, carrying three people 100 kilometers. At least a dozen teams in several countries are attempting to win it.

First space tourist

American financier Dennis Tito, 60, (above) became the world's first space tourist in May, 2001 on board a Soyuz TM-34 rocket for a ten-day mission at the International Space Station. He had paid $20 million for the experience. The cash-strapped Russians have insisted this is the only way they can continue.

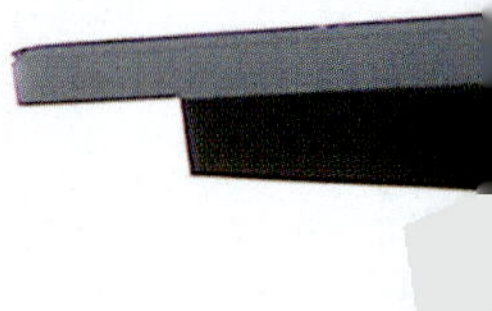

The winner is...

In late 2001, in one of the most important aircraft production battles in military aviation history, Lockheed Martin's design for the Joint Strike Fighter (right) was selected over Boeing's design (above) to become one of the western world's principal new combat planes. The initial $200 billion contract for the F-35 is for some 3,000 planes. The conventional take-off and landing model will replace the USAF's F-16 and A-10, and the US Navy's F/A-X programme; while the VSTOL version will replace the US Marine Corps AV-8B and F/A-18A/C/D and Britain's RAF and Navy Harriers. The JSF project is so big many companies, including Boeing, are expected to share in its production. Canada and several other countries are also participating in it. Entering service about 2010 the aircraft will share 70-90 per cent common parts, be supersonic and carry significantly more weapons over a greater range than the Harriers.

The future?

Europe's Airbus is building the A-380, a 550-seat, double-decker airliner (right, top and below), forecasting airway congestion between airport hubs will get worse and require fewer, larger planes. Boeing, however, is offering the Sonic Cruiser (above) designed to fly at just below the speed of sound and carry only about 220 passengers. It believes demand will be for travel directly and more quickly between regional airports rather than changing planes at hub airports.

A380
A380

Microlight world record

Britons Brian Milton and Keith Reynolds fly a microlight over Hong Kong in 1998 in an attempt to re-create the spirit of aviation's pioneers by being the first to fly such a machine around the world. The frail microlight, a machine the Wright brothers might have instantly recognised, exposed Milton and Reynolds to cold, wind and rain. Flying at 60mph they had aimed to go around the world within 80 days – like Phileas Fogg in Jules Verne's famous story. But they were chased by a MiG 21 over Syria, refused permission to cross into China, and Reynolds gave up his seat to a Russian navigator to get Milton to Alaska. Milton's lone return to England after 120 days established a world record for a type of machine increasingly considered the future for individual aviation.

Balloon challenge

With virtually all the aircraft records already broken, by the end of the 20th century only ballooning seemed to remain as a great aviation challenge. In December, 1998 British aviation entrepreneur Richard Branson and Swedish balloonist Per Lindstrand set off from Morocco in a competition to be the first to circumnavigate the earth in a balloon (above). Their attempt failed over the Pacific, but on 20 March, 1999, the team of Britain's Brian Jones and Switzerland's Bertrand Piccard finally succeeded in the Breitling Orbiter 3.

Free as the wind
In the golden silence of the skies glider pilots and balloonists feel the pleasure of flight as subtle winds and thermals and simple hot air carry them high into the air on the eve of the centenary of manned flight.

Index

Picture acknowledgements

This book was produced by gettyimages Publishing Projects. We would like to thank the following for their assistance.

Agence France Presse Martin Fueger/DPA 378-9; Nabil Ismail 374, 375; Yuri Kochetkov 434; Mike Nelson 394-5; Steffen Schmidt 433; Stf 421 **Airbus Industrie** 439t **Peter Almond** 430b **The Aviation Picture Library** 345, 348-9, 358, 377, Terry Joint 442, Airbus Industrie 382; EADS 376 **The Boeing Company** 189t, 189b, 272b, 310-11t, 438 **Chrysalis Images** 322, 368 **Denver International Airport** 420 **US Dept. of Defense** DoD DVIC MARB CA 364t, 364-5b, 365t **Robert Hunt Library** 203, 211, 231, 340-1 **Lockheed Martin Corporation** 388-9t **Ministry of Defence** Royal Air Force 384; Royal Navy 369 **National Photo Collection/The State of Israel** Sa'ar Ya'acov 352, 353 **Picture Partnership** Colin Edwards 440 **Reuters** Sean Adair 424, 425; Peter Andrews 399; Yannis Behrakis 428-9; The Boeing Company 390-1, 436; Greg Bos 393; Larry Chan 418-19; Dept. of Defense 392, 431m, 431b; Kieran Doherty 381; Jose Miguel Gomez 403; Nikolai Ignatiev 373; Pawel Kopczynski 431t; Lockheed Martin Corporation 437; Muzammil Pasha 400; Efrain Patino 401; Oleg Popov 430t; Raytheon 388-9t; Royal Navy 408; Stringer 416-17, 426; Rob Taggart 380; US Air Force 406, 407, 427; US Navy 404-5; Jason Webb 396-7; Ed Wray 412-13 **The Vulcan Operating Company** 383

All other images in this book are from Getty Images collections including the following which have further attributions:
Slim Aarons 337; **Agence France Presse** 306-7; **Airbus Industrie** 439b; **T Boccon-Gibod/Gamma Liaison** 441; **Mike Fiala** 432; **Peter L Gould** 330-1; **Slava Katamidze Collection** 85, 210; **George de Keerle** 370-1; **Lambert** 204br; **Marine nationale** 411; **Francois Mori** 422-3; **Oleg Nikishin** 435; **NASA** 360, 361; **New York Times Co.** 131, 144-5; **Scott Peterson** 402; **Noel Quidu** 385; **Royal Navy** 410; **Anthony Suau** 387; **Scott Swanson Collection** 222-3; **US Navy** 409, **John Gay/US Navy** 414-5

For information about licensing Getty Images content contact your local Getty Images office or email liz.ihre@gettyimages.com